THE AGE OF DINOSAURS

Consultant: Peter Dodson, Ph.D.

PUBLICATIONS INTERNATIONAL, LTD.

Louis Weber, C.E.O.
Publications International, Ltd.
7373 North Cicero Avenue
Lincolnwood, Illinois 60646

Permission is never granted for commercial purposes.

Manufactured in U.S.A.

8   7   6   5   4   3   2   1

ISBN 0-7853-0443-6

Library of Congress Catalog Card Number 93-84891

Consultant:

Peter Dodson, Ph.D., is a widely recognized expert on dinosaurs and has consulted on many publications, including the *Encyclopedia of Dinosaurs* and *Discover Dinosaurs*, and is coeditor of *The Dinosauria*. He is a professor of anatomy at the School of Veterinary Medicine of the University of Pennsylvania and also teaches courses and lectures on geology, paleontology, and evolution. Dr. Dodson received his doctorate from Yale University.

Contributors:

Brooks Britt: Brooks Britt received his master's degree in geology from Brigham Young University. His credits include working at the Tyrell Museum of Paleontology in Alberta, Canada. His specialty is large theropods.
Kenneth Carpenter: Kenneth Carpenter received his undergraduate degree from the University of Colorado. His credits include acting as senior preparator at the Denver Museum of Natural History. He has mounted and overseen the mounting of more than 50 fossil skeletons. Most of his research concerns ankylosaurs and tyrannosaurs.
Catherine A. Forster: Catherine A. Forster earned her master's degree in geology from the University of Pennsylvania. Her credits include acting as collections manager of paleontology at the Field Museum of Natural History in Chicago. Ornithopods and ceratopsians are her specialty.
David D. Gillette, Ph.D.: David D. Gillette received his doctoral degree in geology from Southern Methodist University. His credits include acting as state paleontologist of Utah. He was also coeditor of *Dinosaur Tracks and Traces*.
Mark A. Norell, Ph.D.: Mark A. Norell received his doctorate in biology from Yale University. His credits include acting as assistant curator of vertebrate paleontology at the American Museum of Natural History. He is a specialist in fossil reptiles and in evolutionary and systematic theory. He has also participated in field projects in South America and Africa.
George Olshevsky: George Olshevsky received his master's degree in computers from the University of Toronto. He has published more than 50 books on a variety of subjects. His credits include publishing a dinosaur newsletter, *Archosaurian Articulations*, and editing *Dinosauria* magazine.
J. Michael Parrish, Ph.D.: J. Michael Parrish earned his doctorate in anatomy from the University of Chicago. His credits include acting as an assistant professor at Northern Illinois University and coediting *Hunteria*, a paleontology journal.
David B. Weishampel, Ph.D.: David B. Weishampel received his master's degree from the University of Toronto and his doctoral degree in geology from the University of Pennsylvania. His credits include acting as associate professor at Johns Hopkins University School of Medicine.

TABLE OF CONTENTS

INTRODUCTION
THE ADVENTURE BEGINS

For much of the Mesozoic Era, a geologic age covering about 180 million years, dinosaurs roamed the earth and achieved dominance over other animal groups. The era was divided into three periods: Triassic, Jurassic, and Cretaceous.

The distant thundering sounds signal the arrival of some of the largest animals ever to have walked the earth. A herd of *Brachiosaurus* arrive to trample every plant they do not eat. The sounds also let a watchful *Allosaurus* know that its next meal may be arriving.

It isn't hard to imagine this scene, even though dinosaurs have been extinct for millions of years. They have captured our imagination, and we delight in learning more and more about them.

The Age of Dinosaurs has been written to lead you into the exciting world of dinosaur research. Start your adventure at the beginning of the book when humans first found dinosaur bones. Early discoveries of skeletons were baffling and probably gave rise to legends of dragons. Find out how explorers came to believe that dinosaurs existed and their early ideas about them.

Dinosaurs pose a special problem for scientists who classify them. With little information and few animals represented, trying to find clues about the dinosaur family tree has turned scientists into detectives. Learn what makes dinosaurs different from other reptiles and what makes them similar.

After learning about the background and history of dinosaurs, you will be introduced to 125 dinosaurs. The first group lived in the Late Triassic and Early Jurassic Periods when there was just one continent, called Pangaea, and no flowering plants. The Middle and Late Jurassic saw the rise of dinosaurs. Sauropods such as *Apatosaurus* and *Brachiosaurus* became the dominant land animals. The Early Cretaceous was a time of further change. Flowering plants appeared, and dinosaurs became more diverse. By the Late Cretaceous, they could truly be said to rule the earth.

And then the dinosaurs died out. The reasons are not understood, but scientists are studying the available information. Every day, they get closer to solving the riddle.

So, begin your adventure in the world of dinosaurs. *The Age of Dinosaurs* will be your travel guide and companion!

CHAPTER 1
DISCOVERING DINOSAURS

Dinosaur bones have been firing our imagination for hundreds of years. People in the Middle Ages found huge bones that were probably fossils of dinosaurs and large aquatic reptiles, which may have inspired the legends of dragons and giants.

Since those first discoveries, scientists have been trying to complete the picture of dinosaurs and their lives. All we have left of these amazing creatures is their fossil bones. Paleontologists—scientists who study fossils—continue their search. Each discovery reveals a little more about dinosaurs, and each new dinosaur tells us, not only about that animal, but also about its place in evolution.

Early Dinosaur Discoveries

Fossil dinosaur bones and teeth continued to turn up in England following the late 17th-century discovery of *Megalosaurus* ("giant lizard") in Oxfordshire. (British scientists considered them large, overgrown lizards.) In a fossil-hunting expedition in the British countryside in 1822, Mary Ann Woodhouse Mantell found fossil teeth that showed for the first time that reptiles of very large size existed long ago. She gave them to her husband, a physician. A worker at the Hunterian Museum in London noticed their resemblance to an iguana's teeth, though they were much larger. French comparative anatomist Baron Georges Cuvier agreed that they were from a giant plant-eating reptile. In 1825, Dr. Gideon Mantell described the teeth and other bones he had found as a new reptilian genus, *Iguanodon*. He later named his second dinosaur *Hylaeosaurus* in 1833. In

Dr. Gideon Algernon Mantell, a 19th-century British physician who described fossil remains of a new reptilian genus, *Iguanodon*.

A dinosaur footprint in Dinosaur State Park in Rocky Hill, Connecticut. Fossils are all that is left of dinosaurs, but they reveal much about the animals' history and habits.

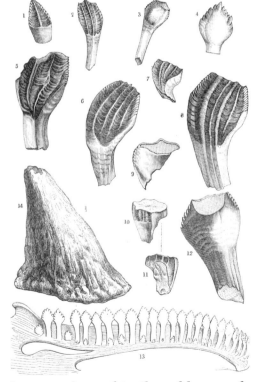

An artist's rendering of *Iguanodon* that appeared in the May 18, 1895, issue of *The Illustrated London News*. The drawing was based on a reconstructed model on display at London's Natural History Museum.

A comparison of teeth and bones of *Iguanodon* and an iguana. The dinosaur's teeth were thought to resemble those of the lizard.

1842, Richard Owen invented the name dinosaur, meaning "terrible lizard."

Harry Govier Seeley, an amateur paleontologist, made a major contribution to dinosaur classification in 1887. He concluded that there were two groups of dinosaurs: the saurischians, with a lizardlike pelvis, and the ornithischians, with a birdlike pelvis. With few changes, scientists still use this classification system today.

America's Dinosaur Rush

In 1858, William Parker Foulke made an amazing discovery while in Haddonfield, New Jersey. John E. Hopkins had found some large spinal bones on his property two decades before. Souvenir hunters had carried off many of the bones, but Foulke and his workers found the rest of the skeleton. The headless specimen was described as a new dinosaur and named *Hadrosaurus foulkii*. It was the first dinosaur skeleton found in North America and one of the best ever discovered until that time. It showed that dinosaurs were not like lizards, but that some walked on their hind legs.

In the late 1860s, two scientists took the spotlight, and their rivalry was to become legendary. Othniel Charles Marsh, nephew of multimillionaire George Peabody, was a Yale graduate. He persuaded his uncle to finance the Peabody Museum at Yale, which supported his field work.

Edward Drinker Cope grew up a Quaker in Philadelphia. He was a child prodigy who, by age 18, had published his

A cartoonist's view of the lighter side of Gideon Mantell's *Iguanodon* discovery. The word *Sawrian* and the theme of the cartoon are a play on words: Mantell described a new genus of reptiles, or saurians. The cartoon appeared in the 1836 edition of *Comic Annual,* published in London.

A SAWRIAN.

first scientific paper. Cope was wealthy, and he later inherited a small fortune that allowed him to pursue his studies.

The two men's rivalry grew out of their ambition to be the greatest paleontologist. By 1866, Cope had described his first dinosaur, the meat-eating *Laelaps aquilunguis.* Marsh had not yet described a dinosaur. He pointed out errors in Cope's 1868 description of the aquatic reptile from Kansas, *Elasmosaurus.* Embarrassed, Cope tried to buy all the copies of his article, but Marsh kept several copies. Cope never forgot the insult.

This ambition to be the first to describe the most spectacular fossils started a "dinosaur rush" that lasted more than a decade. Marsh and Cope hired teams to find new locations, dig for dinosaurs, and send bones to New Haven and Philadelphia for description. Each scrambled to be the first in print. As a result, many dinosaurs received two or more names. For example, Marsh described *Apatosaurus* in 1877 and *Brontosaurus* in 1879. Because they are the same animal, the older name has priority.

As the 19th century was ending, Andrew Carnegie endowed a natural history museum in Pittsburgh. He instructed William J. Holland, the museum director, to find dinosaurs to exhibit. Museum employees Jacob Wortman and Arthur S. Coggeshall found a partial dinosaur skeleton at Sheep Creek, Wyoming. They unearthed another of the same species nearby. Together with bones from related dinosaurs, workers made the two into a spectacular 84-foot-long mounted skeleton that Holland named *Diplodocus carnegii.*

Edward Drinker Cope, a Philadelphia naturalist and paleontologist who discovered some one thousand fossil species. However, he is probably best known for his longstanding rivalry with Othniel Charles Marsh, which came to a head following Cope's faulty 1868 description of the aquatic reptile *Elasmosaurus.*

Dinosaur Discoveries in Canada

In the early 1870s, George Mercer Dawson had found dinosaur bones in Saskatchewan while surveying the Canadian-U.S. boundary. In 1884, Dawson's assistant unearthed part of the skull of a meat-eating dinosaur, later named *Albertosaurus sarcophagus,* in Alberta.

Geologist Thomas Chesmer Weston boated along the Red Deer River to scan the shore for dinosaur fossils. By the turn of the century, Canadian paleontologist Lawrence M. Lambe had collected many dinosaur specimens using the same technique. A few years later, in 1910, Barnum Brown specially outfitted a barge and continued the work of Weston and Lambe.

Charles Hazelius Sternberg's childhood interest in fossil plants led to a job as a fossil collector for Edward Drinker Cope. The Canadian Geological Survey hired Sternberg in 1912 to find dinosaur skeletons and send them to Ottawa. Among the many fossils he found were two duckbilled dinosaur "mummies," preserved with extensive skin impressions.

European Dinosaur Finds

In the late 1870s, coal miners at Bernissart, Belgium, discovered well-preserved dinosaur skeletons more than a thousand feet below ground. Soon the mining company Charbonnage de Bernissart diverted its resources to excavating the skeletons, a task that took several years. Louis Dollo devoted his life to preparing and studying these skeletons, making *Iguanodon* the best-known European dinosaur.

Joseph Leidy, considered the father of American paleontology, with a limb bone from *Hadrosaurus,* which he named and described in 1858.

Othniel Charles Marsh, a 19th-century American naturalist and the guiding force behind Yale University's Peabody Museum. Marsh feuded bitterly with Edward Drinker Cope in a mutual effort to be preeminent in the field of dinosaur research.

An artist's conception of Waterhouse Hawkins's dinosaur display in the Extinct Animals room of Britain's Crystal Palace. The drawing appeared in the December 31, 1853, issue of *The Illustrated London News*.

Europe's premier vertebrate paleontologist at the turn of the century was Friedrich Freiherr von Huene. He was a nobleman—with the time and means to pursue a university education. At the University of Tübingen, he began his long career with a study of the Triassic dinosaurs of Germany. When workers found dinosaur bones at Trossingen, von Huene investigated. The excavations uncovered an enormous bed of *Plateosaurus* bones from the Late Triassic.

Excavating in Asia

The Canadian dinosaur rush was ending. Meanwhile, Henry Fairfield Osborn of the American Museum of Natural History decided the place to look for fossil mammals and the origins of the human race was in the Gobi Desert. In 1922, the museum sent expeditions led by Roy Chapman Andrews and Walter Granger to the Gobi. They found no human remains, but instead located ancient mammal bones and, in one place, dinosaur bones. Heartened by these discoveries, the men returned in 1923. At Shabarakh Usu (now called Bayn Dzak), they identified the first dinosaur eggs ever found, many still arranged in nests. Workers also unearthed skeletons of *Protoceratops andrewsi*, the dinosaur that laid the eggs, finding skeletons from hatchling to adult.

During the 1930s, political turmoil in China prevented further exploration of the Gobi. After World War II, paleontologists from the Soviet Union, led by Ivan Antonovich Efremov, found many fossils in the Nemegt Basin. The huge meat-eater *Tarbosaurus bataar* and the giant plant-eater *Saurolophus angustirostris* were two new finds.

galosaurus display in ...in's Crystal Palace Park. The model, created by Waterhouse Hawkins, was completed in 1854.

China's foremost paleontologist from the 1930s until his death in 1979, Young Chung Chien studied vertebrate paleontology under von Huene in Germany. On returning to China in 1928, he worked for the Geological Survey of China, starting excavations for fossil reptiles in several places. Young wrote up and supervised the mounting of *Lufengosaurus*, discovered in Yunnan Province. It was the first mounted dinosaur skeleton displayed in China. After the Chinese revolution, Young founded the Institute of Vertebrate Paleontology and Paleoanthropology.

Dinosaurs of Africa

W. B. Sattler found the most interesting African dinosaurs in 1907 near Tendaguru Hill, Tanzania (in what was then German East Africa). Berlin paleontologists Werner Janensch and Edwin Hennig led an expedition to the site. From 1909 to 1912, several hundred untrained native workers toiled in the hot, humid climate to excavate the deep pits. They crated and carried by hand thousands of bones, some weighing hundreds of pounds, to the port of Lindi for shipment to Berlin.

In autumn 1912, word of large dinosaur bones found in Egypt's Baharia Oasis reached Ernst Freiherr Stromer von Reichenbach at the University of Munich. The bones belonged to a new meat-eating dinosaur, *Spinosaurus*, which had a six-foot-high sail on its back.

An artist's rendering of prehistoric animals displayed at Britain's Crystal Palace, from *The Land We Live In*, published in the early 1860s.

From the late 19th century until the 1950s, French paleontologists studied the dinosaurs of Morocco, Algeria, and Madagascar. In 1896, Charles Déperet described a long-necked plant-eater, *Lapparentosaurus,* and a large meat-eater, *Majungasaurus,* from Madagascar. After World War II, Albert F. de Lapparent and Réné Lavocat described several new dinosaurs from Morocco and the Sahara Desert.

Dinosaurs from South America

In 1936, an expedition sponsored by the Museum of Comparative Zoology of Harvard University went to the Santa Maria Formation of Rio Grande do Sul. Headed by Llewellyn Ivor Price and Theodore E. White, the prospectors brought back a large collection of fossils, including the partial skeleton of *Staurikosaurus pricei.* Described by Edwin Colbert in 1970, this animal may be the oldest known "true" dinosaur.

Price remained in Brazil, and his work inspired a generation of native South American paleontologists. Among these was Osvaldo A. Reig of the Institut Lillo of Tucuman in Argentina. Reig worked in the Ischigualasto Valley in the San Juan Province, where goat farmer Victorino Herrera found dinosaur remains slightly younger than *Staurikosaurus.* In 1958, Alfred Sherwood Romer and Bryan Patterson uncovered reptile fossils from a period when dinosaurs were establishing themselves.

In 1963, Reig described the primitive dinosaurs *Herrerasaurus* and *Ischisaurus,* which opened a hidden chapter of dinosaur evolution. During the past two decades, José Bonaparte has steadily uncovered new chapters in dinosaur evolution in South America.

A Petrified National Forest landscape. Located in eastern Arizona, this preserve has yielded numerous fossil-bearing rocks that have contributed much to the study of dinosaur evolution.

Excavating fossil remains in the great bone cavern at Maestricht, the Netherlands. The rendering appeared in *The Science Record,* published in 1873 in New York.

11

Dinosaur Research Today

The United States and Canada are home to the most vigorous current dinosaur research. The Tyrrell Museum of Paleontology in Alberta is in the middle of a fertile dinosaur burial ground. Led by Philip J. Currie, Tyrrell researchers have found bone beds that apparently are the remains of dinosaur herds. These can provide information about growth changes, individual differences, disease, and herd structure.

John R. "Jack" Horner has discovered hatchling duckbilled dinosaurs, dinosaur eggs, embryos, and nesting grounds in Montana's Two Medicine Formation. One kind of egg belonged to a duckbilled dinosaur that he and Robert Makela named *Maiasaura*. Another egg was from a small dinosaur that he and David B. Weishampel called *Orodromeus*.

Farther south, in Utah and Colorado, James A. Jensen's work has left an enormous amount of material—enough to fill a warehouse. This includes remains of the immense plant-eaters *Supersaurus*, *Ultrasauros*, and *Dystylosaurus*, the more modest-size *Cathetosaurus*, the meat-eater *Torvosaurus*, and several others.

Arizona, New Mexico, and Texas have dinosaur-bearing rocks from the Late Triassic to the Late Cretaceous. Petrified Forest National Park and its surroundings are of Triassic age and are being studied by Robert A. Long, J. Michael Parrish, and several others researching fossils of North America's oldest known dinosaurs.

A life-size sculpture of the Late Triassic predator *Coelophysis*, on display at the Neville Public Museum, Green Bay, Wisconsin. About nine feet long, *Coelophysis* could rear to a height of perhaps five feet.

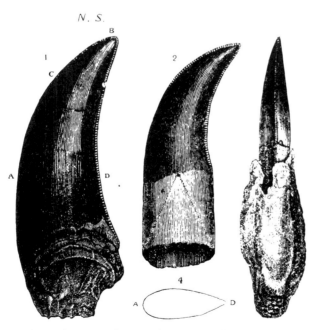

Teeth and jaws of *Megalosaurus*, first described by British clergyman William Buckland in 1824. The fossils were excavated from slate deposits in Oxfordshire. The renderings appeared in Buckland's *Geology and Mineralogy Considered With Reference to Natural Theology*, published in 1836.

Allosaurus fragilis, an agile Late Jurassic predator whose powerful rear legs were built for speed.

In 1947, an American Museum of Natural History field party led by Edwin Colbert discovered an extensive dinosaur burial site at Ghost Ranch, New Mexico, with dozens of skeletons of *Coelophysis bauri* tangled together on the bed of an ancient stream. In 1989, Colbert published his work on *Coelophysis,* making it the best-known Late Triassic predatory dinosaur.

Dinosaur research in China continues. The Cultural Revolution of the 1960s stopped most work, but in the mid-1970s there was an explosion of discoveries. Dinosaurs from the Late Triassic through the Late Cretaceous are now represented in China by excellent material. Joint Canada-China expeditions to remote regions of China have uncovered many treasures soon to be described.

The Polish-Mongolian Paleontological Expeditions of the late 1960s and early '70s returned to the Gobi. Zofia Kielan-Jaworowska led the expeditions, which were rewarded with the discovery of new kinds of dinosaurs and more-complete remains of other known dinosaurs. Inspired by the success of the Polish scientists, the Soviet Union's Academy of Sciences took over field work in the Gobi in the mid-'70s, and since then the Joint Soviet-Mongolian Paleontological Expeditions have unearthed more dinosaur remains. Currently, a series of joint Mongolian-American expeditions is under way involving scientists from the

A museum visitor contemplating the Late Jurassic skeleton of *Stegosaurus,* in this 19th-century drawing.

13

American Museum of Natural History in New York.

Madrid paleontologist José Luis Sanz and his colleagues recently found small and large predators, huge long-necked herbivores (including the new species *Aragosaurus ischiaticus*), small plant-eaters, large plant-eaters *(Iguanodon)*, and armored dinosaurs *(Hylaeosaurus)*. Topping off their work was the discovery at Las Hoyas of a new genus of Early Cretaceous fossil bird, somewhere between the "feathered dinosaur" *Archaeopteryx* and more modern birds.

In the 1970s, French paleontologists discovered interesting Cretaceous African dinosaurs near the Tenere Oasis in the southern Sahara. Besides claws and teeth identified as *Carcharodontosaurus*, they found the skeletons of two new dinosaurs related to *Iguanodon*, one of which was named *Ouranosaurus*.

Discoveries of polar dinosaurs have excited interest. Researchers from the University of California at Berkeley have found abundant dinosaur remains on the North Slope of Alaska, including a skull of the horned dinosaur *Pachyrhinosaurus*. Thomas and Patricia Rich, working on the south coast of Australia, discovered the remains of southern polar dinosaurs. They named one small plant-eater *Leaellynasaura* after their daughter. The frozen continent of Antarctica has also yielded dinosaurs. Since 1987, an armored dinosaur, a two-legged plant-eater, and a meat-eater have been discovered and will soon be named.

Some of the most interesting work does not always involve the discovery of new dinosaurs. For some paleontologists, it is the analysis of fossils gathered years, even decades, ago. We do not know what fossil bone may lead to new insights into dinosaur behavior or evolution. With the continuing work of dedicated scientists, our knowledge of these wonderful creatures increases daily.

The fossilization process. In the top two frames, a ceratopsid has died near a body of water and has been partly covered by sediment. The animal's flesh, muscles, tendons, and other soft parts have started decaying. In the third frame, sediment layers have built up over time, with mineral deposits' filling in pores in the animal's bones. In the final frame, after millions of years have passed, the fossils have become exposed due to erosion, land shift, earthquake, or other natural event.

CHAPTER 2

CLASSIFYING DINOSAURS

Two sauropods beside an Early Cretaceous lake. By the time they left their tracks in the sand, true dinosaurs already had been in existence for more than 100 million years.

Dinosaurs ruled the world during the Mesozoic Era, which is divided into three periods. During the Triassic Period, dinosaur ancestors were evolving. In the Late Triassic, the world saw the first true dinosaurs. In the Jurassic Period, the number of dinosaurs grew. By the Cretaceous Period, many different types of dinosaurs had evolved.

In order to study dinosaurs, scientists use a system of classification called taxonomy. This system groups related animals into species, genera, families, infraorders, suborders, orders, classes, phyla, and kingdoms. Animals are grouped together because of common traits that they inherited from common ancestors.

Because scientists have incomplete information about dinosaurs, groupings may change when new information is found. So each new dinosaur fossil that is discovered could be a key that unlocks new information about dinosaur ancestry.

The Linnaean System of Classification

In the 1750s, Swedish botanist Carl von Linné (better known by the Latin form of his name, Carolus Linnaeus) developed a system to classify all living things. Each creature has two scientific names, a genus and a species. The scientist who first describes a new animal or plant names it. Since Linnaeus began using this system, more than a million species have been named.

In the Linnaean system, similar *species* are grouped into a *genus,* similar *genera* into a *family,* similar *families* into an

CLASSIFICATION CHART

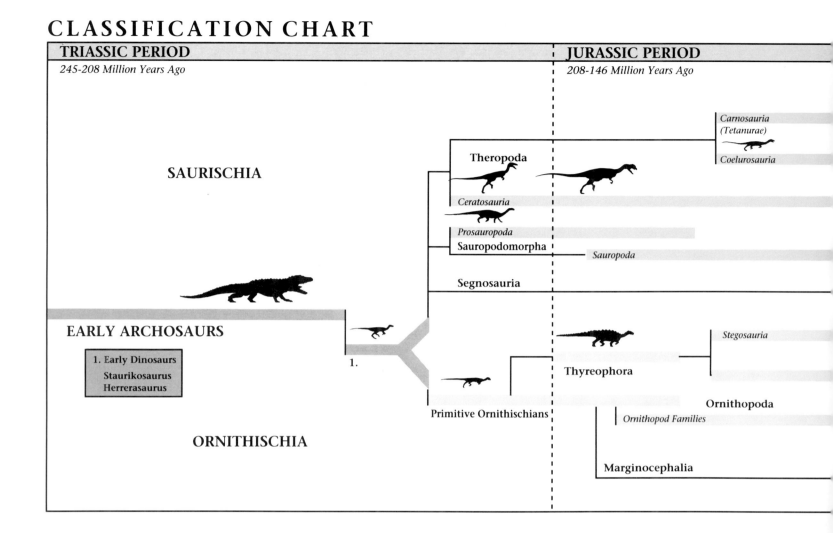

TRIASSIC PERIOD	JURASSIC PERIOD
245-208 Million Years Ago	*208-146 Million Years Ago*

SAURISCHIA

Theropoda

Carnosauria (Tetanurae)

Coelurosauria

Ceratosauria

Prosauropoda

Sauropodomorpha

Sauropoda

Segnosauria

EARLY ARCHOSAURS

1. Early Dinosaurs
Staurikosaurus
Herrerasaurus

1.

Thyreophora

Stegosauria

Primitive Ornithischians

Ornithopoda

Ornithopod Families

ORNITHISCHIA

Marginocephalia

order, similar *orders* into a *class,* similar *classes* into a *phylum,* and similar *phyla* into a *kingdom.* These categories are known as *taxa,* and the study of these classifications is called *taxonomy.*

Dinosaur Relatives

Dinosaurs are classified as reptiles, but all reptiles do not form a single *clade* (a group that includes a common ancestral species and all the species that descended from it). There are two reptilian clades. One includes all living reptiles, dinosaurs, ichthyosaurs, plesiosaurs, and birds (the Sauropsida). The other clade includes the mammals and the extinct mammallike reptiles (the Theropsida).

Crocodilians and birds are more closely related to each other than either is to lizards and snakes. They are part of a smaller sauropsid clade, the Archosauria. Lizards and snakes are in the clade Lepidosauria.

Archosaurs Evolve

Two important evolutionary changes took place among the archosaurs. They changed from sprawling, lizardlike animals to animals that walked with legs held directly under the body. The other change was from being cold-blooded to warm-blooded. Crocodiles are cold-blooded, while birds are warm-blooded. Warm-bloodedness appeared at the same time in the dinosaur-bird clade, so that many dinosaurs may have been warm-blooded.

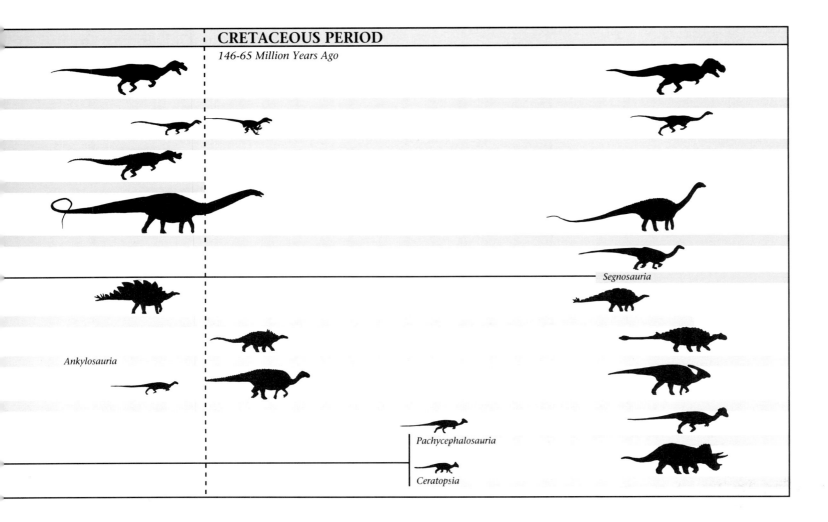

Segnosauria

Ankylosauria

Pachycephalosauria

Ceratopsia

The ancestors of dinosaurs developed a stronger ankle. This kind of ankle occurs in pterosaurs, dinosaurs, and birds.

True Dinosaurs at Last

Harry Govier Seeley split dinosaurs into two groups, the Order Saurischia ("lizard-hipped" dinosaurs) and the Order Ornithischia ("bird-hipped" dinosaurs). Both orders probably had a common ancestor that lived sometime during the Middle Triassic. Birds belong to the saurischian dinosaur clade.

As in all land animals, there were three bones in each side of the pelvis. The left and right *ilia* (singular: *ilium*) firmly gripped the spine in the sacrum. The left and right *pubes* (singular: *pubis*) extended down beneath the ilia. The left and right *ischia* (singular: *ischium*) extended down and back beneath the ilia and behind the pubes. In some dinosaurs, the pubes extended down and forward, as they do in lizards. This is why Seeley called them saurischian or "lizard-hipped" dinosaurs. In other dinosaurs, the pubes extended down and back, running beneath and parallel to the ischia, as in birds. Seeley called these dinosaurs ornithischian or "bird-hipped" dinosaurs.

Some dinosaurs were neither saurischians nor ornithischians. The earliest, most primitive dinosaurs, such as *Staurikosaurus,* fit into neither order.

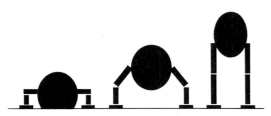

Archosaurs, from which dinosaurs descended, changed from lizardlike animals to ones with legs held directly beneath their bodies.

Dinosaur Families

ORDER: SAURISCHIA
SUBORDER: SAUROPODOMORPHA
Infraorder: Prosauropoda

Prosauropods were widespread and had at least seven families. They lived in the late Triassic until the Early Jurassic. The largest prosauropods, some as long as 40 feet, were straight-limbed dinosaurs that resembled the later sauropods in some ways. All prosauropods were plant-eaters.

Family: Thecodontosauridae. The most primitive prosauropod, *Thecodontosaurus,* was also one of the smallest. It was about six to ten feet long. Like all prosauropods and most sauropods, it had a prominent claw on each front foot and a large claw on each back foot.

Family: Plateosauridae. This is the best-known family of prosauropods, with animals found in Europe, China, and North and South America. They were up to 25 feet long with narrow, long snouts; long necks; powerful front and back limbs; and heavy bodies.

A sketch of a Late Jurassic allosaur feeding upon a sauropod carcass. *Allosaurus* was the most powerful—and most fearsome—meat-eater of its time.

Infraorder: Sauropoda

All sauropods were gigantic four-legged plant-eaters. Like today's elephants, adult sauropods had little fear of predators because of their size. Sauropod skulls were either flat or tapered to a point, and the nostrils were back from the tip of the snout.

Family: Vulcanodontidae. The earliest true sauropod is *Vulcanodon* from the Early Jurassic of Zimbabwe. The only skeleton available is missing the head, neck, and much of the tail. It had a bulky body, and its legs were long and straight. The front limbs were almost as long as the back, and each back foot had five toes.

Three *Plateosaurus engelhardti* at the edge of a Late Triassic forest. The dinosaurs were members of the family Plateosauridae, fossils of which have been found in Europe, Asia, and North and South America.

Family: Barapasauridae. The next most primitive sauropod, *Barapasaurus*, is known from parts of several skeletons from the Early Jurassic of India. It was up to 60 feet long, with a slender body and long neck, tail, and limbs.

Family: Cetiosauridae. Cetiosaurid skulls were blunt and boxlike, with nostrils at the side of the snout. The neck was short, usually with 12 bones. The best-known genus is *Shunosaurus* from the Middle Jurassic of China. It had a small, bony club at the end of its tail.

Family: Brachiosauridae. The front limbs of the brachiosaurids were as long as or longer than the back limbs. This gave the body a backward slope from the neck to the tail. Most brachiosaurids were larger than the cetiosaurids, 80 or more feet long even though they had shorter tails. They were among the heaviest land animals known.

Family: Camarasauridae. In this family, the skull was boxlike. The dinosaurs still had 12 neck bones, and the front limbs were slightly shorter than the back limbs.

Family: Diplodocidae. This family includes some of the best-known sauropods, such as *Apatosaurus* and *Diplodocus*. Diplodocid skulls were long and ended in a spoon-shaped snout; the nostrils were on top of the skull. Small, rod-shaped teeth were in the front of their snout. Diplodocids had long necks, with up to 15 bones.

Family: Titanosauridae. Almost all southern hemisphere sauropods from the Late Cretaceous, and many earlier ones, were titanosaurids. Their limbs were stocky. The spinal bones from the front and middle of the tail were unique—the feature that best distinguishes the family.

SUBORDER: SEGNOSAURIA

Originally classified as a carnivore, *Segnosaurus* had feet with four functional toes and a distinctive pelvis with pubes that pointed back. It is now thought to be a sauropodomorph, but its ancestry is not known.

Skeletal drawings of *Brachiosaurus*. One of the heaviest land animals known, this dinosaur weighed up to 80 tons, or about as much as 12 elephants.

Ceratosaurus nasicornis in a Late Jurassic forest. This powerful theropod was *Allosaurus's* fiercest rival.

SUBORDER: THEROPODA

The theropods include all the predatory dinosaurs. From the smallest dinosaurs to the largest meat-eaters, the theropods included the most different kinds of saurischian dinosaurs of all suborders. These two-legged meat-eaters had clawed feet with no more than three functional toes.

Theropods evolved into two major groups: the Ceratosauria, with flexible tails; and the Tetanurae, with stiff tails.

Infraorder: Ceratosauria

Family: Podokesauridae. The earliest ceratosaurians include *Coelophysis* from the Late Triassic of western North America. This was a small, nimble dinosaur with a long, slender skull and many teeth.

Families: Halticosauridae and Ceratosauridae. *Dilophosaurus* lived during the Early Jurassic. It had a double crest on its head. *Ceratosaurus* was from the Late Jurassic and had a horn on its head. Both were from western North America.

Infraorder: Tetanurae

Family: Compsognathidae. The most primitive tetanuran was *Compsognathus* from the Late Jurassic of Europe. It was the smallest theropod, two to three feet long and lightly built.

Family: Coeluridae. *Ornitholestes* and *Coelurus,* which lived during the Late Jurassic in western North America, were fast-running, lightly built theropods up to three feet tall at the hips and six to ten feet in length.

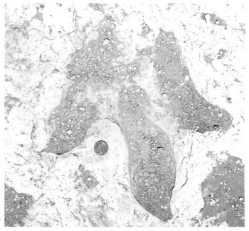

Three-toed theropod tracks from the Kayenta Formation in northeastern Arizona. Theropods were meat-eaters that walked or ran on their hind legs.

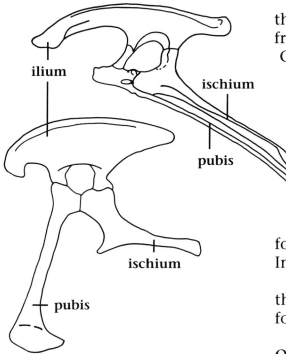

ilium

ischium

ischium

pubis

pubis

Pelvic drawings of a typical ornithischian, or "bird-hipped," dinosaur *(top)* and a saurischian, or "lizard-hipped," dinosaur *(bottom)*.

Family: Allosauridae. This family is typical of the larger Jurassic and Early Cretaceous theropods that·sometimes exceeded 35 feet in length. The biggest allosaurid may have been more than 40 feet long. Allosaurids were slender but dangerous predators.

Family: Tyrannosauridae. Most dramatic of all the theropods were the tyrannosaurids, which probably came from allosauridlike ancestors at the beginning of the Late Cretaceous. Unlike other tetanurans, they had massive bodies and small, two-fingered hands.

Family: Ornithomimidae. Many kinds of small theropods also arose in the Cretaceous. Their skeletons were birdlike. Most had hands and front limbs with powerful striking action.

Family: Dromaeosauridae. The discovery of the sickle-clawed *Deinonychus* supported the idea of a bird-dinosaur relationship and started the debate about dinosaurs' being warm-blooded. Each foot had a large, sickle-shaped claw on the second toe. Interlocking bones stiffened the end of the dinosaur's tail.

Family: Troodontidae. This group of "sickle claw" theropods had large brains and large eyes that faced forward.

ORDER: ORNITHISCHIA

The earliest ornithischian dinosaur was *Pisanosaurus*, a three-foot-long, two-legged plant-eater from the late Middle Triassic of Argentina. All ornithischians were plant-eaters.

Compsognathus attacking two *Archaeopteryx*. *Compsognathus*, which lived during the Late Jurassic, was one of the smallest known dinosaurs. Except for its modified long front legs and feathers, *Archaeopteryx* closely resembled its predator.

Family: Fabrosauridae. All fabrosaurids were small plant-eaters that walked on their two hind legs. The best-known fabrosaurid is *Lesothosaurus,* from the Early Jurassic of southern Africa.

SUBORDER: THYREOPHORA
Infraorder: Stegosauria

Stegosaurs were the main armored dinosaurs of the Jurassic Period; ankylosaurs remained in the background. But ankylosaurs replaced stegosaurs in importance during the Cretaceous Period.

Family: Stegosauridae. These armored dinosaurs may have evolved in China during the Early Jurassic. By the Late Jurassic, besides China *(Tuojiangosaurus),* they were in Europe *(Dacentrurus),* North America *(Stegosaurus),* and Africa *(Kentrosaurus).*

Thescelosaurus neglectus, in a Late Cretaceous forest. One of the last of the hypsilophodontids, it may have witnessed the mass extinction at the end of the Mesozoic Era.

Infraorder: Ankylosauria

During the Early Cretaceous, ankylosaurs replaced stegosaurs in importance. Ankylosaurs were different from stegosaurs—the heads of stegosaurs were long and narrow, but ankylosaurs had short, broad skulls.

Family: Nodosauridae. The more primitive ankylosaurians, including all Jurassic and southern hemisphere genera, belong in this family. Some nodosaurids had large, cone-shaped spines along the neck and shoulders for protection.

Family: Ankylosauridae. This family may have arisen during the Early Cretaceous from a nodosaurid ancestor. Ankylosaurid skulls had horns projecting from the back. All ankylosaurids had massive, bony tail clubs for defense.

Euoplocephalus tutus, from the Late Cretaceous. An ankylosaurid, this dinosaur is thought to have had an especially keen sense of smell.

A pair of *Lagosuchus* perched above a Middle Triassic lake. These archosaurs had advanced ankles and other features suggesting they were closely related to dinosaurs.

SUBORDER: MARGINOCEPHALIA
Infraorder: Pachycephalosauria

The back of the head of these dome-headed dinosaurs was broadened into a shelf that often had bony lumps or short spikes. In one family, Pachycephalosauridae, the bones were raised into a very thick, high dome that was the main feature of the animal's appearance and even grew over the shelf (as in *Stegoceras*). Pachycephalosaurs had broad, chubby bodies and were bipedal plant-eaters.

Infraorder: Ceratopsia

Family: Psittacosauridae. The oldest and most primitive ceratopsians belong in this family of small, bipedal runners such as *Psittacosaurus* from the Early Cretaceous.

Family: Protoceratopsidae. In this family, the back of the skull was expanded into a short, wide frill over the neck.

Family: Ceratopsidae. From cow to elephant size, the four-legged ceratopsids had horns and frills on their heads. They had powerful jaws with hundreds of teeth for slicing tough plants. *Triceratops* had the most powerful jaw muscles of any land animal.

A *Psittacosaurus* skeleton. One of the most primitive ceratopsians, it was also—in the juvenile stage—one of the smallest dinosaurs. The skull of one juvenile fossil found in Asia is only about an inch long.

SUBORDER: ORNITHOPODA

Family: Heterodontosauridae. These small, nimble, two-legged plant-eaters have been found mainly in Early Jurassic rocks of southern Africa. Their teeth were sharp and tusklike in front, but the teeth at the sides of the jaw were made for slicing and chewing plants.

A dinosaur egg from the Late Jurassic Period.

Family: Hypsilophodontidae. This was the most widespread and longest-lived ornithopod family. Hypsilophodontids were small, but had relatively large heads. Their feet were primitive with four functional toes.

Family: Dryosauridae. This short-lived family arose about the same time as the Hypsilophodontidae. Dryosaurids, with small front limbs and heads, were larger and more powerful than hypsilophodontids. They lacked teeth at the front of the snout and instead had a well-developed beak.

Family: Camptosauridae. The Late Jurassic genus *Camptosaurus* from western North America was a chubby, medium-size ornithopod about 15 feet long, with specialized feet and skull.

Family: Iguanodontidae. *Iguanodon* is one of the best-known dinosaurs. This bulky, 30-foot-long ornithopod had a deep, narrow skull; a strong, well-developed pelvis; rows of bony tendons running along its back; a hand in which the thumb had become a sharp spike; and three broad toes, plus an inner toe reduced to a splint.

Family: Hadrosauridae. This family consists of two groups, the Hadrosaurinae and the Lambeosaurinae. They were both duckbilled dinosaurs. The teeth were in 30 to 40 vertical rows, like steps on a moving escalator. As each tooth wore away, it was replaced by the one directly below it.

The duckbilled dinosaurs were perhaps the most advanced of all the dinosaurs. They had excellent hearing, eyesight, voice, and sense of smell. They lived in huge herds and may have migrated seasonally.

A *Parasaurolophus* herd at sunset. Like other duckbilled dinosaurs, *Parasaurolophus* was blessed with good hearing, eyesight, and sense of smell. It also had what seemed like a horn on top of its head—actually a resonating chamber that allowed the animal to make honking noises.

CHAPTER 3
DINOSAUR BEGINNINGS

Triassic and Early Jurassic Periods

Dinosaurs had not arrived as the Triassic Period began. The earth's land and climate were changing, the ancestors of the dinosaurs were evolving, and many plants and animals were becoming extinct. By the Jurassic Period, dinosaurs became rulers of the earth.

At the beginning of the Triassic Period, all land was joined as a single large mass called Pangaea, stretching nearly from pole to pole. The only block of land not joined to Pangaea was southern China and southeast Asia, which collided with the Chinese mainland before the Late Triassic. Throughout the Triassic, Pangaea slowly moved north and turned clockwise. By the middle of the period, the continent was a crescent-shaped mass facing east and centered along the equator.

Dinosaurs did not appear until the Middle to Late Triassic, but many other animals lived in the Triassic world. The therapsids were mammallike reptiles that ruled the earth before dinosaurs. They included dicynodonts, the turtlebilled plant-eaters; and cynodonts, the ancestors of mammals. Also living then were the labyrinthodonts, the late survivors of the early amphibians; and procolophonids, the early plant-eating reptiles.

Archosaurs first appeared in the Late Permian as a minor part of the fauna (dinosaurs, crocodiles, pterosaurs, and

Triassic map.

Riojasaurus, **a Late Triassic prosauropod giant from South America.**

A herd of plateosaurs grazing beside a lake in Late Triassic Germany.

thecodonts all belong to the group Archosauria). Halfway through the Early Triassic, important archosaurs appeared— *Erythrosuchus,* a large meat-eater; and *Euparkeria,* a fast, small meat-eater.

During the Middle Triassic, the ancestors of the dinosaurs were beginning to be important. *Lagosuchus, Pseudolago-suchus,* and *Lagerpeton* were tiny meat-eaters with long legs. They were close to the common ancestors of dinosaurs and pterosaurs, with ankles like later dinosaurs.

The earliest dinosaur remains appeared in rocks that date to the late Middle Triassic or earliest Late Triassic in Brazil and Argentina. They included *Staurikosaurus, Herrerasaurus,* and *Pisanosaurus.* Halfway through the Late Triassic, a major change happened. New reptiles unrelated to dinosaurs appeared, and a number of plant-eating groups, including dicynodonts, disappeared. Where this happened, theropod (meat-eating) dinosaurs appeared in larger numbers.

At the end of the Triassic, a major extinction occurred. Archosaurs became extinct except for dinosaurs, crocodilians, and flying reptiles called pterosaurs. These extinctions probably happened because Pangaea was moving north and breaking apart, causing the climate to change.

In the Early Jurassic, few new major groups of dinosaurs appeared. Instead, most of the existing dinosaurs increased in number and diversity.

An early Jurassic scene. During this time, few dinosaur groups emerged—with the exception of the first sauropods, *Vulcanodon* and *Barapasaurus.*

Two *Coelophysis* in Late Triassic North America. The ten-foot-long predator, a migratory dinosaur, moved in large herds.

ANCHISAURUS

(ANK-ee-SORE-us)

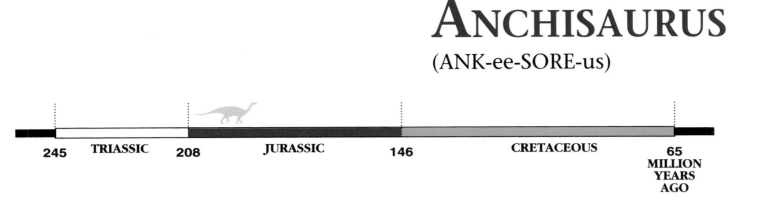

245	TRIASSIC	208	JURASSIC	146	CRETACEOUS	65 MILLION YEARS AGO

*A*nchisaurus was one of the earliest North American dinosaurs described. The first partial skeleton was found in Connecticut in 1818 and was thought to be a human fossil. Scientists recognized *Anchisaurus* as a dinosaur in 1885. Its name means "near reptile," because it seemed so much like its reptilian ancestors.

Anchisaurus was different from most other prosauropods because its front and back feet were long and narrow. Peter Galton and Michael Cluver classified *Anchisaurus* and *Thecodontosaurus* as narrow-footed prosauropods, separating them from the prosauropods that had broader feet. *Anchisaurus* and *Thecodontosaurus* are among the smallest prosauropods, so the difference in foot shape may be because of their size.

The skull of *Anchisaurus* was lightly built and triangular when viewed from the side. Other prosauropods, such as *Plateosaurus,* had more rectangular skulls with flatter snouts. *Anchisaurus* also differed from other prosauropods because the jaw joint was level with the lower jaw, rather than below the tooth row. *Anchisaurus* had blunt, diamond-shaped teeth that were less tightly packed and fewer than in other prosauropods. All this shows that *Anchisaurus* probably ate soft, easily chewed plants.

Period:
Early Jurassic

Order, Suborder, Family:
Saurischia, Sauropodomorpha, Anchisauridae

Location:
North America (United States)

Length:
6½ feet (2 meters)

Anchisaurus, for nearly 70 years mistaken for a human fossil, had blunt teeth probably suited to eating soft plants.

Scientists recognized *Anchisaurus* as a dinosaur in 1885. Its name means "near reptile," because it seemed so much like its reptilian ancestors.

COELOPHYSIS
(SEE-loh-FIE-sis)

A mass of tangled bodies rolled with a flood, sliding over trees that had fallen into the muddy waters. The rains stopped and hundreds of carcasses of *Coelophysis*, the nimble predator of the Late Triassic, settled into the mud. Some skeletons were complete, some were torn apart, but all went to the bottom of the stream. Two hundred million years later, at Ghost Ranch in northern New Mexico, paleontologists unearthed a treasure trove of dinosaur skeletons. They were all from one group devastated by a flood in the Late Triassic. The animals here ranged from hatchlings to adults more than two meters long.

The body of *Coelophysis* had a long slender tail and jaws filled with dozens of knife-edged teeth. Its head was large, with a pointed snout and large eyes.

Besides the skeletons from Ghost Ranch, *Coelophysis* has also been found in the Painted Desert of Arizona. The petrified logs found there, many longer than 100 feet, show us what the forests looked like when these dinosaurs ran about.

Coelophysis is the oldest dinosaur in the world known from many complete skeletons. The name *Coelophysis* means "hollow condition," referring to the hollow bones of the legs. They were built much like birds' bones for minimum weight and maximum strength. *Coelophysis bauri* is the only species known. *Coelurus* was an early name used for some of the original bones, which were mistakenly given several names.

In the rib cages of two adults from Ghost Ranch are the skeletons of young *Coelophysis*. They are too large and well developed to be unborn babies. This may have been cannibalism—one individual of a species eating another—and the prey was swallowed whole.

Rela-tives of *Coelophysis* include *Podokesaurus; Halticosaurus* and *Procompsognathus* from Germany; and *Syntarsus* from Zimbabwe and Arizona.

Unlike the giant predatory dinosaurs of later times, the Triassic theropod dinosaurs were also prey. Their enemies included the enormous phytosaurs, which weighed a ton or more, and the active and powerful rauisuchid predators such as *Postosuchus*. These adversaries dominated the Triassic landscape, both on land and in the water.

A *Coelophysis bauri* skeleton from Ghost Ranch in northern New Mexico. It is among the oldest dinosaurs in North America.

Coelophysis's rear legs were strong and agile. It had feet with three long toes and one short one, and it was quick to leap away from larger predators. The front legs were small and were not used for walking. They were more likely used to gather food.

A *Coelophysis* skeletal drawing. The dinosaur lived in large herds, something that does not happen in today's world. Although grazing animals such as wildebeest or caribou live in herds in our modern world, no predators live in large groups.

The body of *Coelophysis* had a long slender tail and jaws filled with dozens of knife-edged teeth.

Period:		
Late Triassic		

Period:
Late Triassic

Order, Suborder, Family:
Saurischia, Theropoda, Podokesauridae

Location:
North America (United States)

Length:
10 feet (3 meters)

A master of ambush, *Coelophysis* was perhaps a fish-eater, a 50-pound predator that lived along streams, moving through ferns and horsetails, always on guard against enemies. It also ate insects, lizardlike reptiles, and other small dinosaurs.

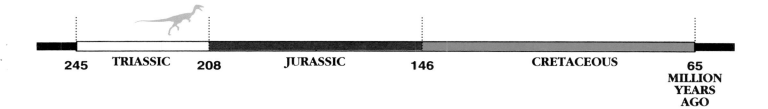

245 **TRIASSIC** **208** **JURASSIC** **146** **CRETACEOUS** **65 MILLION YEARS AGO**

29

DILOPHOSAURUS
(die-LOH-foh-SORE-us)

Period:
Early Jurassic
Order, Suborder, Family:
Saurischia, Theropoda, Podokesauridae
Location:
North America (United States)
Length:
20 feet (6 meters)

The terror of the Early Jurassic, *Dilophosaurus* was one of the earliest large theropods. It weighed nearly 1,000 pounds and was the dominant predator of its time.

The teeth in the front of the snout were long and slender, probably useful for plucking and nipping at the flesh of its prey. The cheek teeth were long, pointed, and blade-shaped, typical of meat-eating dinosaurs. The lower jaw was slender, and the bones of the neck were not large. The muscles of the jaws and neck were not as powerful or as large as in later theropods. *Dilophosaurus* did not overpower its prey like its Late Jurassic relatives *Allosaurus* and *Ceratosaurus*. It slashed and tore at the flesh of its victim until it fell.

Balanced on its large hips, *Dilophosaurus* was fast and agile. The front legs were small and not often used for running. Three of the four fingers on the hands had claws that gripped and tore at the prey when it was feeding. The rear legs were long and made for running. The rear feet had three toes that were covered with claws. The muscular tail was as long as the front part of the body, and it helped balance the animal when it was moving.

Dilophosaurus had two parallel crests on the top of its head, from the tip of the snout to the top of the skull between the eyes. There was a thin bone in each crest. The animal's name means "two-ridged reptile," after these crests. Other theropods had horns and crests, but none were as prominent.

These crests made *Dilophosaurus*'s head look rounder and larger; this may have helped it when competing for food or territory. Some paleontologists have suggested that this early theropod was social and that it settled disputes with rivals by displaying its crest, which may have shown the animal's social position. Perhaps the dominant males, with control of territory and females, had the largest and most colorful heads and crests. These crests would have also warned away challengers.

Before the skull of *Dilophosaurus* was discovered, this early theropod was confused with the European meat-eater *Megalosaurus*. *Dilophosaurus* lived with the smaller theropods and the primitive ornithischian dinosaurs in the Kayenta Formation of Arizona. *Dilophosaurus* ate the large, plant-eating prosauropod dinosaurs and any other prey it could capture.

Dilophosaurus, with its parallel crests. Scientists have various theories about the nature of the crests, each of which had a thin bone (thus the name, "two-ridged reptile").

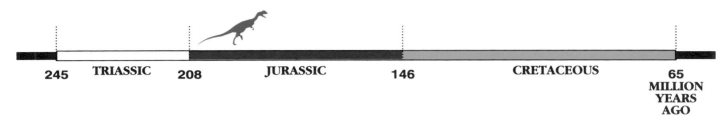

| 245 | TRIASSIC | 208 | JURASSIC | 146 | CRETACEOUS | 65 MILLION YEARS AGO |

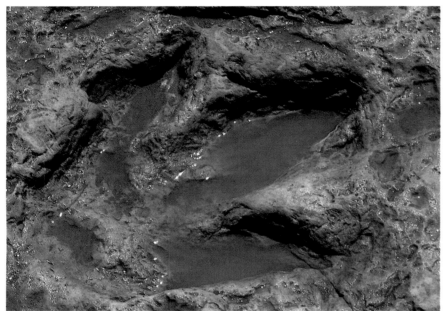

A fossil footprint of *Dilophosaurus wetherilli* in the Painted Desert of Arizona. The dinosaur's strong rear legs ended in three-toed feet covered with claws.

The terror of the Early Jurassic, *Dilophosaurus* was one of the earliest large theropods. It weighed nearly 1,000 pounds and was the dominant predator of its time.

The dinosaur may have been a social animal, and colorful crests could have indicated social standing. Perhaps, too, the crests helped males assert control over territory and females by warning away rivals.

31

PLATEOSAURUS

(PLAT-ee-oh-SORE-us)

Plateosaurus is the best known of the early dinosaurs. It was among the largest of the Triassic dinosaurs, reaching an adult height of 15 feet when it stood on its back legs. An adult probably weighed close to a ton.

One distinctive feature of *Plateosaurus* and of all prosauropods is their hands. They had small fingers and a huge thumb with a large claw. This claw may have been used for plucking leaves from high branches, for digging roots, or for fighting. The front limbs were shorter than the back ones, but the dinosaur often walked on all four limbs.

The head of *Plateosaurus* was small. It was longer and flatter than the heads of the later sauropods. Its lower jaw curved down, as did the front of its snout; this made

The back legs of *Plateosaurus* were long and thick. The feet had five toes, with the fifth (outer) toe very small. The pelvis was short and massive, and the dinosaur had large, powerful leg muscles.

the animal look as if it was smiling. It had simple, leaf-shaped teeth. They did not change shape from the front to the back of the jaw.

To help its digestion, *Plateosaurus* may have had "gastroliths." These were stones in the stomach (swallowed by the dinosaur) that rubbed against one another and the food, slicing and grinding the leaves and plants to help digestion.

Plateosaurus fed on high vegetation, such as the leaves of tree ferns. Prosauropods probably browsed in high tree branches much as sauropods did later in the Mesozoic and as giraffes do today. Scientists A. W. Crompton and John Attridge noted that prosauropods were unique among Triassic plant-eaters because they had lightly built skulls, which suggests that they probably ate soft plants. To chew tougher branches and leaves, an animal's skull and jaw must be heavily built.

Dinosaurs did not become common until near the end of the Triassic. In Late Triassic European quarries, *Plateosaurus* is the most common vertebrate fossil.

In Halberstadt and Trossingen, Germany, workers found mass graves of *Plateosaurus* and have removed dozens of skeletons from each. Trossingen contains one layer that has a herd of plateosaurs killed by a catastrophic event, possibly a flood. Paleontologist David Weishampel suggested that the reason *Plateosaurus* is so common in other layers of the quarry is that it was the most abundant animal at the time.

A complete *Plateosaurus* skull. Teeth in the upper jaw fitted between pairs of teeth in the lower jaw. This allowed *Plateosaurus*, a plant-eater, to chop leaves as it ate. *Plateosaurus* probably had fleshy cheeks that stored food to keep it from falling out as the animal ate.

| **Period:** |
| Late Triassic |
| **Order, Suborder, Family:** |
| Saurischia, Sauropodomorpha, Plateosauridae |
| **Location:** |
| Europe (Germany, France, Switzerland) |
| **Length:** |
| 23 feet (7 meters) |

One distinctive feature of *Plateosaurus* and of all prosauropods is their hands. They had small fingers and a huge thumb with a large claw. This claw may have been used for plucking leaves from high branches, for digging roots, or for fighting. The front limbs were shorter than the back ones, but the dinosaur often walked on all four limbs.

Two plateosaurs. The prosauropods were unique among Triassic dinosaurs because they had long necks. This allowed them to reach leaves and branches high off the ground. *Plateosaurus* could reach even higher when it reared up on its back limbs.

A muscle drawing of *Plateosaurus engelhardti.*

| 245 | **TRIASSIC** | 208 | **JURASSIC** | 146 | **CRETACEOUS** | 65 **MILLION YEARS AGO** |

BARAPASAURUS

(bah-RAP-ah-SORE-us)

Named for a word meaning "big leg" in a local dialect in central India, *Barapasaurus* is the oldest known sauropod dinosaur. Early Jurassic sauropods are rare. The sauropods may have evolved from a prosauropod ancestor. *Barapasaurus* proves that even the earliest sauropods were giants.

Barapasaurus lived at the same time as the last prosauropod dinosaurs. In some ways, *Barapasaurus* was similar to the prosauropods, but in other ways it was quite advanced. Its spinal bones and large size prove it belongs to the sauropods. *Barapasaurus* was almost as large as *Diplodocus,* a Late Jurassic sauropod of North America. *Barapasaurus* was probably a plant-eater, using its long neck to gather leaves from treetops.

The only mounted dinosaur skeleton in India, where dinosaurs are rare, is *Barapasaurus.* The skeleton includes mostly leg and spinal bones. The skull was not found, but several teeth were located nearby. The teeth were spoon-shaped; these may have been the front teeth that were used for cropping vegetation. The chewing teeth were likely larger and flattened for crushing food.

Relatives of *Barapasaurus* include *Cetiosaurus* from northern Africa and England, *Patagosaurus* and *Volkheimeria* from southern South America, *Amygdalodon* from Argentina, *Lapparentosaurus* from Madagascar, and possibly *Rhoetosaurus* from Australia. Descendants of *Barapasaurus* and its relatives dominated the Jurassic world.

Period:
Early Jurassic
Order, Suborder, Family:
Saurischia, Sauropodomorpha, Cetiosauridae
Location:
Asia (India)
Length:
49½ feet (15 meters)

Barapasaurus, India's only mounted dino-saur skeleton, was probably a plant-eater. In local dialect, its name means "big leg."

The teeth were spoon-shaped; these may have been the front teeth that were used for cropping vegetation. The chewing teeth were likely larger and flattened for crushing food.

245	**TRIASSIC**	208	**JURASSIC**	146	**CRETACEOUS**	65 **MILLION YEARS AGO**

EUSKELOSAURUS
(you-SKELL-oh-SORE-us)

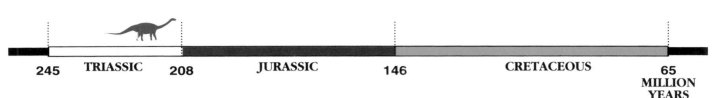

| 245 | TRIASSIC | 208 | JURASSIC | 146 | CRETACEOUS | 65 MILLION YEARS AGO |

*E*uskelosaurus was a prosauropod giant. Unfortunately, we do not have a complete skeleton—only limb and spinal bones. Thomas H. Huxley named it in 1866—*Euskelosaurus* means "true-limbed reptile."

Euskelosaurus had huge limbs. Their length and massive build and the overall size of the dinosaur support the idea that *Euskelosaurus* was among the most sauropodlike of the prosauropods.

The dinosaur's thigh bone had a twisted shaft. Paleontologist Jacques van Heerden suggested that this twisting meant that the back legs of *Euskelosaurus* were "bow-legged." If van Heerden is correct, this would be unusual for a dinosaur. All other dinosaurs' thighs were directly under their bodies.

The habits of *Euskelosaurus* remain a mystery because the head, hands, and feet are unknown. Based on the parts of the skeleton we have, the dinosaur appears to have been a large, slow-moving plant-eater somewhat similar to *Apatosaurus*. However, we will not know for certain until scientists find better fossils.

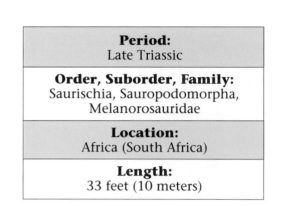

Period:
Late Triassic
Order, Suborder, Family:
Saurischia, Sauropodomorpha, Melanorosauridae
Location:
Africa (South Africa)
Length:
33 feet (10 meters)

Euskelosaurus had huge limbs. Their length and massive build and the overall size of the dinosaur support the idea that *Euskelosaurus* was among the most sauropodlike of the prosauropods.

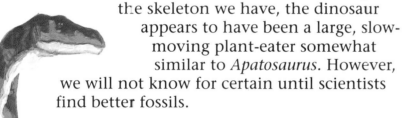

Euskelosaurus brownii. Called the "true-limbed reptile," it was also big-limbed. *Euskelosaurus's* thigh bone with its twisted shaft may mean the animal was bow-legged.

35

HERRERASAURUS

(huh-RARE-uh-SORE-us)

This carnivore weighed perhaps 500 pounds and stood about four feet tall at the shoulders. It reached 18 feet in length. *Herrerasaurus* was one of the dominant theropods of the Late Triassic of South America. The skull was armed with sharp, daggerlike teeth. *Herrerasaurus* was a meat-eater that had muscular jaws and claws on the front feet.

Herrerasaurus lived around 228 million years ago, when dinosaurs first appeared in the Triassic Period but were not yet dominant. These earliest dinosaurs were all small predators, like *Coelophysis* in North America.

Herrerasaurus represents the roots of dinosaur evolution. It was a theropod dinosaur, a distant ancestor of *Tyrannosaurus*. This early and very primitive dinosaur had four toes on its back feet. This separates it from other meat-eating dinosaurs, which had three toes. It was advanced in other ways, however. Its pelvis and back bones were similar to the advanced theropods of the Jurassic and Cretaceous Periods.

Staurikosaurus, which some paleontologists place in the same family with *Herrerasaurus*, also had four toes on its back feet. *Herrerasaurus* was slightly more advanced in the hips, however.

Period:
Late Triassic
Family:
Herrerasauridae
Location:
South America (Argentina)
Length:
18 feet (5.5 meters)

With a hip structure more advanced than another early dinosaur, *Staurikosaurus, Herrerasaurus* is thought to be a distant relative of the tyrannosaurids.

Herrerasaurus ischigualastensis, a medium-size meat-eater that lived about 228 million years ago, was discovered by—and named after—Argentine goat farmer Victorino Herrera.

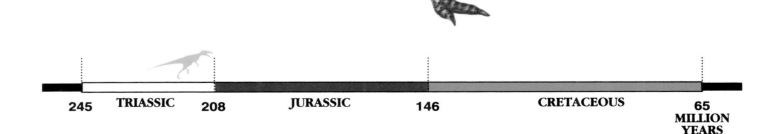

| 245 | TRIASSIC | 208 | JURASSIC | 146 | CRETACEOUS | 65 MILLION YEARS AGO |

HETERODONTOSAURUS
(HET-ur-oh-DONT-oh-SORE-us)

245	TRIASSIC	208	JURASSIC	146	CRETACEOUS	65 MILLION YEARS AGO

*H*eterodontosaurus was a small plant-eater. The front limbs were muscular, and the joints of the elbow and wrist show that it used its front limbs for grasping and not for walking. The fourth and fifth fingers were small and probably opposed the first three, forming a grasping hand for grabbing food.

The pelvis of *Heterodontosaurus* was long and narrow. The pubis was almost like that found in later ornithischians. The back legs were long, with the lower leg and foot longer than the thigh. The ankle bones and all but one of the bones in the foot joined together, as in birds.

Most dinosaurs had only one kind of tooth, but this dinosaur had three kinds. Its name means "different-toothed reptile." The front teeth were small and were on the sides of the beak on both the upper and lower jaws. These teeth were probably used to chop leaves and stems. The back teeth (cheek teeth) were tall and squared off, and the upper and lower teeth met at an angle for grinding plants. A third type—large, paired tusks—were in front of the cheek teeth.

Heterodontosaurus probably had fleshy cheeks. These would have kept food from falling out of the mouth during chewing.

Most dinosaurs had only one kind of tooth, but this dinosaur had three kinds. Its name means "different-toothed reptile."

Period:
Early Jurassic
Order, Suborder, Family:
Ornithischia, Ornithopoda, Heterodontosauridae
Location:
Africa (South Africa)
Length:
3 feet (90 centimeters)

Heterodontosaurus tucki. The dinosaur's front and back limbs were similar to later ornithischians (except for its unusual joined ankle and foot bones). Also, the angled cheek teeth, the jaw, and the fleshy cheeks were all features of later ornithischians.

LUFENGOSAURUS

(loo-FUNG-oh-SORE-us)

Lufengosaurus was a close relative of *Plateosaurus*. It was about the same size, but it lived earlier. There are many fossils of this dinosaur from the Lufeng basin of Yunnan Province in southwestern China. It is one of several prosauropods known from the Early Jurassic Lufeng Formation, along with some early crocodiles, mammals, and therapsids.

Although there are complete skeletons of this dinosaur, it is not fully described. The skull of *Lufengosaurus* was long and flat, and it had a small bump on its snout just above the nostril. It had a long neck.

Its teeth were bladelike with crowns that were wider at the bottom. The teeth were widely spaced. Its diet is unknown. It probably ate plants, but its teeth were sharp, and it may have also eaten small animals.

The front legs of *Lufengosaurus* were shorter than the powerful back legs. The animal walked on all fours, but probably rose on its back legs to feed on tall plants. The hands of *Lufengosaurus* had a large thumb with a claw, used for getting food and maybe as a weapon.

A skeleton of *Lufengosaurus* was the first dinosaur mounted in the People's Republic of China. It appeared on a Chinese postage stamp when the skeleton went on display.

Period:
Early Jurassic
Order, Suborder, Family:
Saurischia, Sauropodomorpha, Plateosauridae
Location:
Asia (People's Republic of China)
Length:
20 feet (6 meters)

A skeletal drawing of *Lufengosaurus,* the first dinosaur to be displayed in China. It had a long neck and a long, flat skull with a bump on the snout.

A quadrupedal dinosaur—that is, one that walked on all fours—*Lufengosaurus huenei* had short front legs and stout hind legs. It probably was a plant-eater, but scientists think it also may have eaten small animals.

(Lufengosaurus) appeared on a Chinese postage stamp when the skeleton went on display.

245	TRIASSIC	208	JURASSIC	146	CRETACEOUS	65 MILLION YEARS AGO

MASSOSPONDYLUS
(MASS-oh-SPON-di-luss)

| 245 | TRIASSIC | 208 | JURASSIC | 146 | CRETACEOUS | 65 MILLION YEARS AGO |

*M*assospondylus was a medium-size prosauropod. Its name means "massive spinal bone." There are many skeletons of this animal, including several good skulls from South Africa and one good skeleton from the Kayenta Formation of Arizona. The Kayenta skull is about 25 percent bigger than the largest African specimen. Some of the smaller skulls from South Africa are probably juveniles. They had taller, narrower skulls with bigger eye sockets. This shows that skull proportions changed as the dinosaur grew.

The skull of *Massospondylus* was shallower and shorter than that of *Plateosaurus,* with more changes along the tooth row. The front teeth were round, but the back teeth were more oval. Teeth in the lower row were shorter than in the upper. The teeth and jaws of *Massospondylus* were made for a diet of plants.

Several of the South African specimens of *Massospondylus* had rounded stones in or near their rib cages. These appear to have been used to grind food in the stomach, much like the gizzards of birds.

The proportions of the skeleton of *Massospondylus* suggest that it could rear up and possibly walk on its hind legs. *Massospondylus* had an enlarged thumb claw that could have been used as a grooming tool or to dig for or grasp food.

A skeletal drawing of *Massospondylus.* The dinosaur was named in 1854 by English anatomist Sir Richard Owen.

| **Period:** |
| Early Jurassic |
| **Order, Suborder, Family:** |
| Saurischia, Sauropodomorpha, Plateosauridae |
| **Location:** |
| Africa (South Africa, Zimbabwe), North America (United States) |
| **Length:** |
| 13 feet (4 meters) |

Two *Massospondylus* in a desert landscape. Many prosauropod dinosaurs like *Massospondylus* have been found in Late Triassic and Early Jurassic formations in Europe, Asia, and South America.

MUSSAURUS

(MOOSE-sore-us)

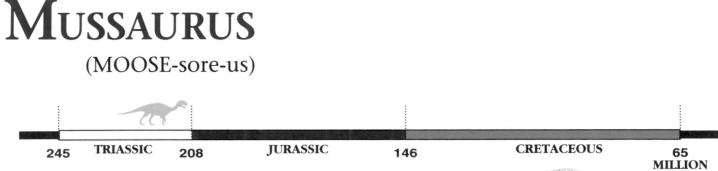

| 245 | TRIASSIC | 208 | JURASSIC | 146 | CRETACEOUS | 65 MILLION YEARS AGO |

In the mid-1970s, José Bonaparte led an expedition to the Late Triassic El Tranquilo Formation of southern Argentina. There was a nearly complete skeleton about six inches long without its tail. Although this and other specimens were all juveniles, *Mussaurus* ("mouse reptile") is now very well known because the skeletons were complete. More recently, adult skeletons have been found in the same formation.

Mussaurus was immediately recognizable as a prosauropod in the limb and pelvic area, although the hatchlings had short necks and high, short skulls with large eye sockets. The younger animals had different body proportions than did adults.

Mussaurus might be halfway between the prosauropods and sauropods and might be closely related to the ancestors of sauropods. It is also possible that many sauropod features in juveniles are only proportional differences in skeleton between juveniles and adults.

Mussaurus provided important information about dinosaur social behavior. The hatchlings and eggs in the nest show that these dinosaurs laid eggs. Several eggs in the nest show that prosauropods laid more than one egg at a time. Now that juveniles, immature specimens, and adults of *Mussaurus* are known, scientists can study changes among animals of different ages.

Mussaurus, or "mouse reptile."

A *Mussaurus patagonicus* juvenile. *Mussaurus* may prove to be an important animal in the search for a relationship between prosauropods and sauropods.

Mussaurus provided important information about dinosaur social behavior.

Period:
Late Triassic
Order, Suborder, Family:
Saurischia, Sauropodomorpha, Plateosauridae
Location:
South America (Argentina)
Length:
Estimated 10 feet (3 meters)

RIOJASAURUS
(ree-OH-hah-SORE-us)

Riojasaurus and the related South African genus *Melanorosaurus* are prosauropod giants. Both had a number of features in common with the sauropod dinosaurs. These include spinal bones with hollow spaces and dense, massive limb bones.

Riojasaurus walked on all four legs; it could not rear up on its back legs like some of its dinosaur relatives, such as *Plateosaurus.* Its long, massive body needed to be supported by all four limbs. The front limbs of *Riojasaurus* were nearly as long as its back limbs. It had four spinal bones connecting the pelvis to the trunk.

Some scientists consider *Riojasaurus* and *Melanorosaurus* the closest relatives of the sauropods because of their large size and some features of their limbs. Recently, however, Peter Galton and Paul Sereno have argued that the prosauropods and sauropods had a common ancestor sometime in the Triassic. If this is true, the similarities probably exist because both groups were large animals. More will be known when newly discovered material of *Riojasaurus,* including at least one good skull, is described by scientists.

Riojasaurus was an early and short-lived prosauropod that lived at the end of the Late Triassic. In the Early Jurassic, smaller animals such as *Anchisaurus* became the dominant prosauropods, and the first sauropods, *Vulcanodon* and *Barapasaurus,* appeared.

Period:
Late Triassic
Order, Suborder, Family:
Saurischia, Sauropodomorpha, Melanorosauridae
Location:
South America (Argentina)
Length:
33 feet (10 meters)

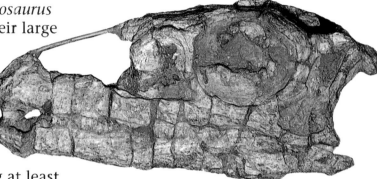

A skull of the prosauropod giant, *Riojasaurus.*

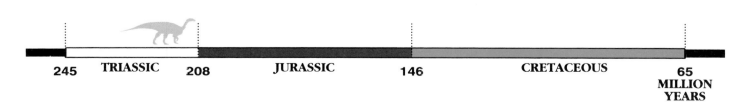

Riojasaurus incertus. A quadruped, it differs from many prosauropods in that it had four spinal bones— rather than three— connecting trunk and pelvis.

| 245 | **TRIASSIC** | 208 | **JURASSIC** | 146 | **CRETACEOUS** | 65 **MILLION YEARS AGO** |

SCELIDOSAURUS

(skee-LIE-doh-SORE-us)

Scelidosaurus was a plant-eating dinosaur that grew to about 13 feet. It was a heavily built animal. The most unusual feature of *Scelidosaurus* was that it had many bony plates in the skin of its back and rib cage. These plates were not like the high, thin plates that extended from the backbone of stegosaurian dinosaurs. These plates were similar to the bony armor of ankylosaurs.

The large head of *Scelidosaurus* was equipped with somewhat simple, leaf-shaped teeth. It probably ate a mixture of leaves from shrubs and low-lying branches, but may also have fed on succulent fruits and even eaten insects when it was a hatchling and juvenile.

The legs of *Scelidosaurus* were stout and the feet had four toes. The size and shape of the legs show that *Scelidosaurus* walked on all four legs. The tail of *Scelidosaurus* was long compared to most other ornithischian dinosaurs. It also had armor in the skin of the tail.

Recently, new discoveries of *Scelidosaurus* have been found from the area where the original fossils were unearthed. These include skull bones and impressions of small, rounded scales in the skin.

Scelidosaurus, whose name means "limb reptile," is one of the most primitive armored dinosaurs. It was first discovered in the middle of the 19th century in Early Jurassic rocks from southern England.

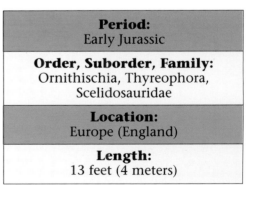

A skeletal drawing of *Scelidosaurus oweni*. From the teeth and the shape of the skeleton, scientists know that *Scelidosaurus* was a plant-eating animal.

| **Period:** |
| Early Jurassic |
| **Order, Suborder, Family:** |
| Ornithischia, Thyreophora, Scelidosauridae |
| **Location:** |
| Europe (England) |
| **Length:** |
| 13 feet (4 meters) |

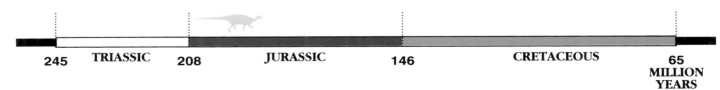

| 245 | TRIASSIC | 208 | JURASSIC | 146 | CRETACEOUS | 65 MILLION YEARS AGO |

SCUTELLOSAURUS

(skoo-TELL-oh-SORE-us)

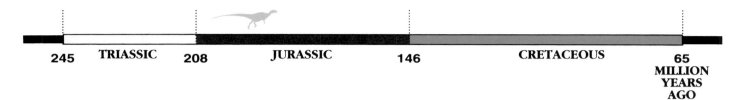

| 245 | TRIASSIC | 208 | JURASSIC | 146 | CRETACEOUS | 65 MILLION YEARS AGO |

Period:	Early Jurassic
Order, Suborder, Family:	Ornithischia, Ornithopoda, Fabrosauridae
Location:	North America (United States)
Length:	4 feet (1.2 meters)

One of the earliest ornithischian dinosaurs, *Scutellosaurus* is unusual because it was a two-legged dinosaur with armor. About the size of a collie dog, it had slender skull bones and jaws with powerful muscles. The animal had bladelike front teeth that were used for grabbing or nipping at plants. The teeth on the sides of the jaws were pointed, with tiny ridges on the sides that helped slice food.

The front legs were strong for a two-legged animal. The front feet were also large, showing that it often used its front legs for walking. The tail was strong and probably useful not only when it was moving but also in defending itself against predators. The rear legs were well suited for walking upright, and the long tail balanced the animal.

The unusual feature of *Scutellosaurus,* or "shield reptile," is the presence of several hundred small bony plates in the skin probably covering the neck, back, ribs, and tail. They are similar to the bony plates of alligators and crocodiles.

Because of the armor and the overall form of the body, some paleontologists think *Scutellosaurus* was the ancestor of the later large, armored dinosaurs, such as *Stegosaurus* and *Ankylosaurus.*

Scutellosaurus, or "shield reptile," had armored plates that probably protected its pelvic area and skin from coarse vegetation, much like modern-day armadillos.

The animal had bladelike front teeth that were used for grabbing or nipping at plants. The teeth on the sides of the jaws were pointed, with tiny ridges on the sides that helped slice food.

43

SELLOSAURUS
(SELL-oh-SORE-us)

*S*ellosaurus skeletons have been found in the middle Late Triassic beds in Nordwürttemberg, Germany. These beds are slightly older than those where the skeletons of *Plateosaurus* were found. Although Frederich von Huene described *Sellosaurus* in 1907, confusion has surrounded the relationships of this dinosaur, *Plateosaurus,* and other members of the Plateosauridae. Using paleontological detective work, Peter Galton cleared up the mystery. By careful comparisons, Galton has shown that *Sellosaurus* appeared first and was the only prosauropod in the middle Late Triassic.

Sellosaurus was similar to *Plateosaurus,* but smaller. Its teeth changed more from the front to the back of the jaw. This and other details of the skull may mean that *Sellosaurus* had a slightly different, softer plant diet than *Plateosaurus.* The two animals are similar in most details and are close relatives. An animal very much like *Sellosaurus* was probably the ancestor of *Plateosaurus.*

The sharp contrast between the small number of *Sellosaurus* specimens and the abundance of *Plateosaurus* shows the sudden spread of prosauropods at the end of the Triassic and in the Early Jurassic.

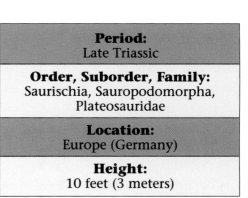

Period:
Late Triassic
Order, Suborder, Family:
Saurischia, Sauropodomorpha, Plateosauridae
Location:
Europe (Germany)
Height:
10 feet (3 meters)

The sharp contrast between the small number of *Sellosaurus* specimens and the abundance of *Plateosaurus* shows the sudden spread of prosauropods at the end of the Triassic and in the Early Jurassic.

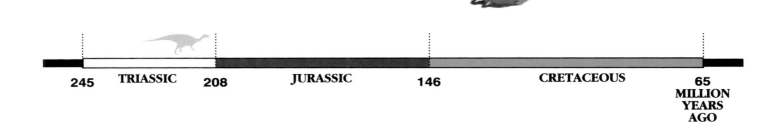

Sellosaurus gracilis. Much confusion has surrounded *Sellosaurus's* relationship to *Plateosaurus.* Careful research has shown, however, that *Sellosaurus* was followed by the other dinosaur, but did not overlap in time with it.

| 245 | **TRIASSIC** | 208 | **JURASSIC** | 146 | **CRETACEOUS** | 65 **MILLION YEARS AGO** |

STAURIKOSAURUS
(store-ICK-oh-SORE-us)

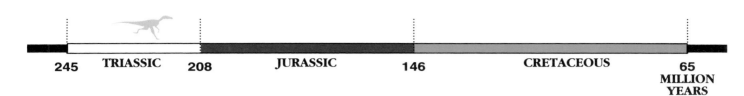

| 245 | TRIASSIC | 208 | JURASSIC | 146 | CRETACEOUS | 65 MILLION YEARS AGO |

Staurikosaurus is the earliest dinosaur known. The first skeleton came from the Santa Maria Formation of southern Brazil, which is either latest Middle Triassic or earliest Late Triassic. Its name means "cross reptile," after the Southern Cross, a group of stars best seen in the southern skies.

Staurikosaurus was a two-legged dinosaur, although its back legs and pelvis were not well suited to rapid motion. It had five fingers and five toes, which is a primitive feature. The third finger and toe were the longest. The teeth were curved backward, with jagged edges like a steak knife. It was a meat-eater.

Paleontologists have differing ideas about *Staurikosaurus*. This animal either represents primitive dinosaurs that share a common ancestry with all later dinosaurs or is the earliest member of the Saurischia. *Staurikosaurus* differs from other dinosaurs in the structure of the pelvis, lower legs, and ankles.

Staurikosaurus fossils are not common, but they occur in formations where other land vertebrates, such as aetosaurs and rauisuchids, are common. The reason there are few *Staurikosaurus* fossils in these deposits could be because there were few animals, or perhaps they lived in areas where bones rarely fossilized, such as forests.

| **Period:** |
| Late Middle or Early Late Triassic |
| **Family:** |
| Staurikosauridae |
| **Location:** |
| South America (Brazil, Argentina) |
| **Length:** |
| 6½ feet (2 meters) |

A skeletal drawing of *Staurikosaurus*. A partial skeleton from Arizona, nicknamed Gertie, is currently being studied.

A pair of *Staurikosaurus*. There are few fossil remains of this early dinosaur. Like *Coelophysis*, which lived at about the same time, *Staurikosaurus* disappeared by the Early Jurassic.

45

SYNTARSUS

(sin-TAR-sus)

The agile, fleet-footed meat-eater *Syntarsus* was a small predator. It strongly resembles *Coelophysis* from the Late Triassic of Arizona and New Mexico. The bones of the head and jaws were large compared to *Coelophysis*, and the neck was somewhat shorter. *Syntarsus* was sturdier and slightly larger than *Coelophysis*.

The dinosaur's jaws had many small, bladelike teeth for slicing the flesh of its prey. *Syntarsus* probably ate other small reptiles and fish, and it lived along stream courses in herds. Like *Coelophysis*, a large number of *Syntarsus* remains were found in one area, suggesting that they were social animals.

The animal is named for its fused ankle, or tarsus, joint. Its name means "fused tarsus." This ankle gave it greater speed and endurance when running. It would have needed to run quickly to escape predators.

The front legs were quite small and weak, while the rear legs were stout. The tail was large and probably stiff. The limbs' proportions suggest that *Syntarsus* may have moved by hopping, similar to rabbits or kangaroos. This hopping produces abrupt and unpredictable movements in order to escape predators.

The many skeletons from Africa have two body forms: Males were small and slender, while females were larger and more robust. There were also many more females than males.

Period:
Early Jurassic
Order, Suborder, Family:
Saurischia, Theropoda, Podokesauridae
Location:
Africa (Zimbabwe), North America (United States)
Length:
10 feet (3 meters)

A muscle drawing of *Syntarsus*.

The dinosaur's jaws had many small, bladelike teeth for slicing the flesh of its prey. *Syntarsus* probably ate other small reptiles and fish. . . .

Syntarsus kayentakatae. The bones of *Syntarsus* had so many bone cells and blood vessels that they resembled the bones of birds and mammals. This may mean that *Syntarsus* was a warm-blooded dinosaur.

	245	TRIASSIC	208	JURASSIC	146	CRETACEOUS	65 MILLION YEARS AGO

YUNNANOSAURUS

(yu-NAN-oh-SORE-us)

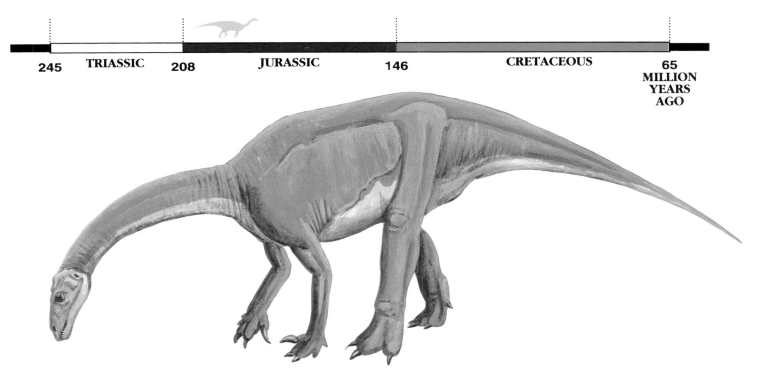

| 245 | TRIASSIC | 208 | JURASSIC | 146 | CRETACEOUS | 65 MILLION YEARS AGO |

Yunnanosaurus huangi. **In 1951, Young Chung Chien described a second species, *Yunnanosaurus robustus,* a larger, more heavily built animal. After further study, however, it was learned that the larger specimen was simply the adult of the first species.**

A pair of *Yunnanosaurus* beside an Early Jurassic stream.

| **Period:** |
| Early Jurassic |
| **Order, Suborder, Family:** |
| Saurischia, Sauropodomorpha, Yunnanosauridae |
| **Location:** |
| Asia (People's Republic of China) |
| **Length:** |
| 18–25 feet (5.5–7.5 meters) |

*Y*unnanosaurus was a long-necked, four-legged plant-eater that lumbered through the fernlike vegetation of southern China. It had more than 60 spoon-shaped teeth that sharpened themselves by wearing against each other as the animal fed. This was different from the dinosaur's prosauropod relatives, which had leaf-shaped teeth that wore away against the food that was being chewed. The advanced teeth of *Yunnanosaurus* were somewhat like those of the sauropods, the dominant plant-eaters of the Middle and Late Jurassic periods. The teeth of *Yunnanosaurus* probably evolved separately, though, because the rest of its anatomy was not like the sauropods'.

One of the last of the prosauropods, *Yunnanosaurus* also had a skull with a different shape than its earlier relatives.

Yunnanosaurus was named after the Yunnan Province in the People's Republic of China. In 1939, Young Chung Chien discovered several partial skeletons, and Wang Tsun Yi excavated them. Young described *Yunnanosaurus huangi* in 1942; it was a smallish, lightly built animal.

The remains of as many as 20 individuals of various sizes have been unearthed in the Lufeng Formation in the People's Republic of China. After these skeletons have been studied and described, we should soon know more about this dinosaur.

47

CHAPTER 4
DINOSAURS TAKE OVER

Middle and Late Jurassic Period

By the Middle and Late Jurassic, dinosaurs had taken over the world. There were herds of *Apatosaurus;* each adult was as large as five or six elephants. *Allosaurus,* a two-ton meat-eater, waited in the bushes for its next meal; it needed to eat often to fill its appetite. Bony plates protected slow, steady *Stegosaurus,* which had little fear of predators. The land quaked with dinosaur footsteps.

Pangaea continued to break apart during the Jurassic. It was splitting both north-south and east-west, and land masses were beginning to resemble today's shapes. The Tethys Sea separated the southern land mass, called Gondwanaland, from the northern mass, called Laurasia. Laurasia consisted of North America, Europe, and Asia. From time to time, there may have been land connections between Gondwanaland and Laurasia because of sea level changes.

The northern part of the Atlantic Ocean began to open during the Jurassic, and it separated Laurasia into eastern (Europe and Asia) and western (North America) land masses. There were probably land bridges connecting them across the north when the sea level dropped.

All this changed the way the ocean waters flowed. Cold ocean currents in the southern hemisphere produced temperate climates in what is now South America, southern

Jurassic map.

***Diplodocus*, a North American sauropod that arose during the Late Jurassic.**

Africa, Antarctica, India (an island off eastern Africa), and Australia. The rest of the world was also warm and moist, and the Triassic deserts began shrinking and were gone by the Late Jurassic. There were no polar ice caps.

The mammallike reptiles that were so important in the Triassic were gone by the Middle Jurassic, as were the rauisuchids and other archosaurs. Late in the Jurassic, small lizards, frogs, and salamanders crawled around under the cover of plants. Turtles, though not large, had also appeared.

This landscape provided food, forage, and home for the dinosaurs. As the land became greener in the Late Jurassic, the sauropods gained the advantage. They reached from China to North America to Africa. From the earliest sauropods in the Early and Middle Jurassic—and the meat-eaters that evolved with them—arose the giants of the Late Jurassic. *Apatosauros, Diplodocus, Brachiosaurus, Ultrasauros, Supersaurus, Allosaurus,* and *Ceratosaurus* roamed the earth. Others, like the ornithopods *Stegosaurus* and *Camptosaurus,* survived the great deserts of the Middle Jurassic and added to the growing group of dinosaurs.

More and more kinds of dinosaurs were evolving in the Middle and Late Jurassic Period, but the most important dinosaurs of this time were the sauropods. These immense creatures roamed the earth searching for food to sustain their enormous size. These animals were the most successful dinosaurs, with their numbers growing rapidly.

The theropod *Ceratosaurus* (right) prepares to attack the armored plant-eater *Stegosaurus.*

A pair of nesting *Archaeopteryx.*

Apatosaurus, a sauropod that came to dominance during the Jurassic.

49

ALLOSAURUS

(AL-oh-SORE-us)

Allosaurus means "other reptile," a strange name for the most powerful, fearsome, and deadly dinosaur of the Late Jurassic. *Allosaurus* was the main enemy of the giant sauropods, including *Apatosaurus* and *Diplodocus,* even though the plant-eaters weighed much more than their two-ton predator.

At the Cleveland-Lloyd Dinosaur Quarry in Utah, more than 10,000 bones have been excavated since 1927, and about half are *Allosaurus* bones. The allosaurs from this quarry include all ages. There were animals from only a little over a foot in length to those more than 40 feet long and weighing more than two tons. Other allosaurs were estimated at more than 49 feet in length and weighing almost three tons. Two other predators at the Cleveland-Lloyd Quarry were *Stokesosaurus* and *Marshosaurus;* both were small and slender. A third predator found was *Ceratosaurus,* a medium-size theropod. All these animals were rare compared to *Allosaurus.*

The powerful skull of *Allosaurus* was the perfect meat-eating machine. The jaws were large and massive, with jagged-edged teeth as well designed for cutting meat as a steak knife. *Allosaurus* probably overpowered its prey and used its massive jaw muscles, its powerful neck and head, and its daggerlike teeth to kill and eat its prey. Allosaurs were widespread. They left their broken teeth near the bodies of many animals, showing where they had been.

The front limbs of *Allosaurus* were short but strong, and the hands had three fingers each. Each finger had a sharply curved and pointed claw. It probably used its front limbs to capture prey and grab the flesh when feeding. The rear legs were large and powerful, built for both speed and agility.

Allosaurus probably ate any animal it could ambush or overpower. The tremendous size of the sauropods was probably an obstacle for even the largest *Allosaurus.* Like lions today, *Allosaurus* was probably opportunistic, attacking old or weak animals when possible and stealing carcasses from other predators.

The most common species is *Allosaurus fragilis.* A close relative of *Allosaurus* is *Acrocanthosaurus* from the Early Cretaceous in North America.

Period:
Late Jurassic
Order, Suborder, Family:
Saurischia, Theropoda, Allosauridae
Location:
North America
Length:
35-40 feet (11-12 meters)

With a skull that served as a perfect meat-eating machine, strong front limbs, and legs built for speed, no wonder *Allosaurus* was the most fearsome theropod of the Late Jurassic.

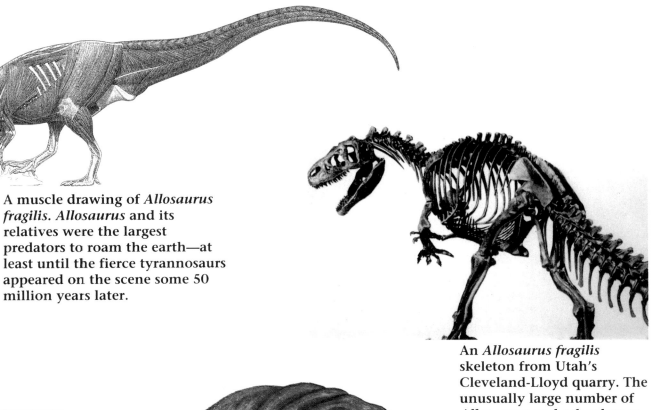

A muscle drawing of *Allosaurus fragilis*. *Allosaurus* and its relatives were the largest predators to roam the earth—at least until the fierce tyrannosaurs appeared on the scene some 50 million years later.

An *Allosaurus fragilis* skeleton from Utah's Cleveland-Lloyd quarry. The unusually large number of *Allosaurus* and other bones found there suggest that both predator and prey got trapped in the quarry at the same time.

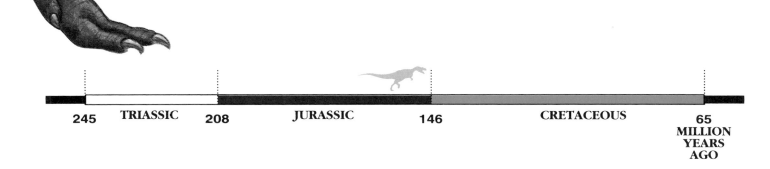

Allosaurus was the main enemy of the giant sauropods, including *Apatosaurus* and *Diplodocus,* even though the plant-eaters weighed much more than their two-ton predator.

| 245 | TRIASSIC | 208 | JURASSIC | 146 | CRETACEOUS | 65 MILLION YEARS AGO |

APATOSAURUS

(uh-PAT-oh-SORE-us)

*A*patosaurus moved constantly, feeding day and night. When it walked, the ground thundered: Each adult weighed as much as five adult elephants. The literal meaning of its name, "deceitful reptile," is difficult to apply to this giant. With a shoulder height of 12 feet, a length of around 70 feet, and a weight of 30 tons or more, this peaceful plant-eater could neither hide nor disappear into the background.

Like its close relatives *Diplodocus, Barosaurus,* and *Supersaurus,* this sauropod had a small head, a long, slender neck, and a deep, heavy midsection. It also had legs built like pillars and a long, heavy tail that ended in a slender whiplash.

The neck of *Apatosaurus* was small near the head, but the base of the neck near the body was huge, and the neck bones were long and massive. *Apatosaurus* could have reached as high as 35 feet to browse at the tops of trees.

The bones of the middle part of the body were huge. This is where *Apatosaurus* carried at least half its weight, about 15 tons. Its ribs were long and straight. The bones of the spine were huge but had hollow spaces that made them lighter, yet no less strong.

Apatosaurus was taller at the hips than the shoulders; the height of a full-grown animal's hips was about 15 feet. Its tail bones near the front of its tail were also huge, and all of the tail bones had tall spines where muscles attached for holding the tail off the ground. The tail probably weighed several tons and may have balanced the animal when it walked. The length of the tail, around 30 feet to the tip, helped distribute the dinosaur's weight.

Apatosaurus, like the other sauropods, was a plant-eater. Paleontologists argue about how it could eat enough to keep a 30-ton body alive. The skull and jaws seem too small to keep enough food coming in. Also, the dominant plants of the time, the evergreens, may not have been nutritious enough for these giants. One adaptation that aided their digestion was gastroliths or "stomach stones" that were in their digestive tracts. The dinosaurs swallowed small stones that then helped grind up the plants in their stomachs.

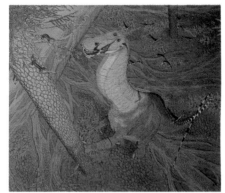

An *Apatosaurus louisae* browsing for food. At around 70 feet in length, it could reach as high as 35 feet into treetops.

Period:	Late Jurassic
Order, Suborder, Family:	Saurischia, Sauropodomorpha, Diplodocidae
Location:	North America
Length:	70 feet (21 meters)

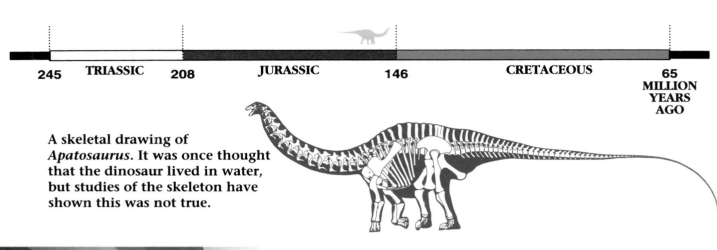

| 245 | TRIASSIC | 208 | JURASSIC | 146 | CRETACEOUS | 65 MILLION YEARS AGO |

A skeletal drawing of *Apatosaurus*. It was once thought that the dinosaur lived in water, but studies of the skeleton have shown this was not true.

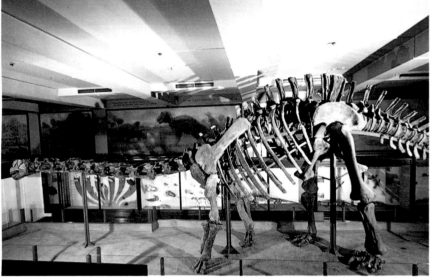

An *Apatosaurus* skeleton. The dinosaur's legs were straight and massive, like an elephant's, with short, stubby toes. The toes of the front feet were blunt, except for the inner toe, which had a claw that pointed inward. The rear feet had three claws and looked a little like cow's hooves.

With a shoulder height of 12 feet, a length of around 70 feet, and a weight of 30 tons or more, this peaceful plant-eater could neither hide nor disappear into the background.

Apatosaurus. For years, artists and scientist visualized the dinosaur as dragging its tail on the ground. But what is far more likely is that *Apatosaurus* carried its tail well off the ground, letting it sway gracefully as it helped the animal keep its balance.

53

ARCHAEOPTERYX
(AR-kee-OP-ter-iks)

Period:
Late Jurassic

Class, Subclass, Order, Family:
Aves, Archaeornithes, Archaeopterygiformes, Archaeopterygidae

Location:
Europe (Germany)

Wingspread:
2 feet (60 centimeters)

*A*rchaeopteryx lithographica is known from only six fossils—they may be the most famous fossils in the world. Its name means "ancient wing from the printing stone." It was found in limestone quarries in the Solnhofen region of southern Germany, and limestone has been used for decades in lithography, or printing. Fossils found in this fine-grained Late Jurassic lagoon sediment are very well preserved. The fossils also include the rarest fossil find: feathers.

The skeleton of *Archaeopteryx* was so much like the small meat-eating dinosaurs that several specimens were catalogued as dinosaurs in museum collections. After closer study, scientists discovered the faint impression of feathers. Because of the feathers, *Archaeopteryx* is usually classified as a primitive bird, but some paleontologists place *Archaeopteryx* with the dinosaurs.

Except for the feathers, *Archaeopteryx* is much like the tiny dinosaur *Compsognathus* and other small coelurosaurs. The biggest difference is that the long front legs of *Archaeopteryx* were modified to support feathers. The three front toes had claws, and the jaws had pointed teeth.

Archaeopteryx had a long, stiff tail, much like the tail of the small theropod dinosaurs; tail feathers were attached to it. In all other birds, the tail bones are almost entirely lost. A stubby leftover tail, called the pygostyle, is where the tail feathers attach.

There is still much argument about the place of *Archaeopteryx* in evolution. The relationship to dinosaurs is strong, and *Archaeopteryx* could be the "missing link" between birds and the small meat-eating dinosaurs. This would mean that we see the descendants of dinosaurs every time we see a modern bird.

Was it a bird? Was it a dinosaur? Paleontologists often find themselves on opposite sides of the evolutionary fence when it comes to *Archaeopteryx*.

Three Ways to Fly

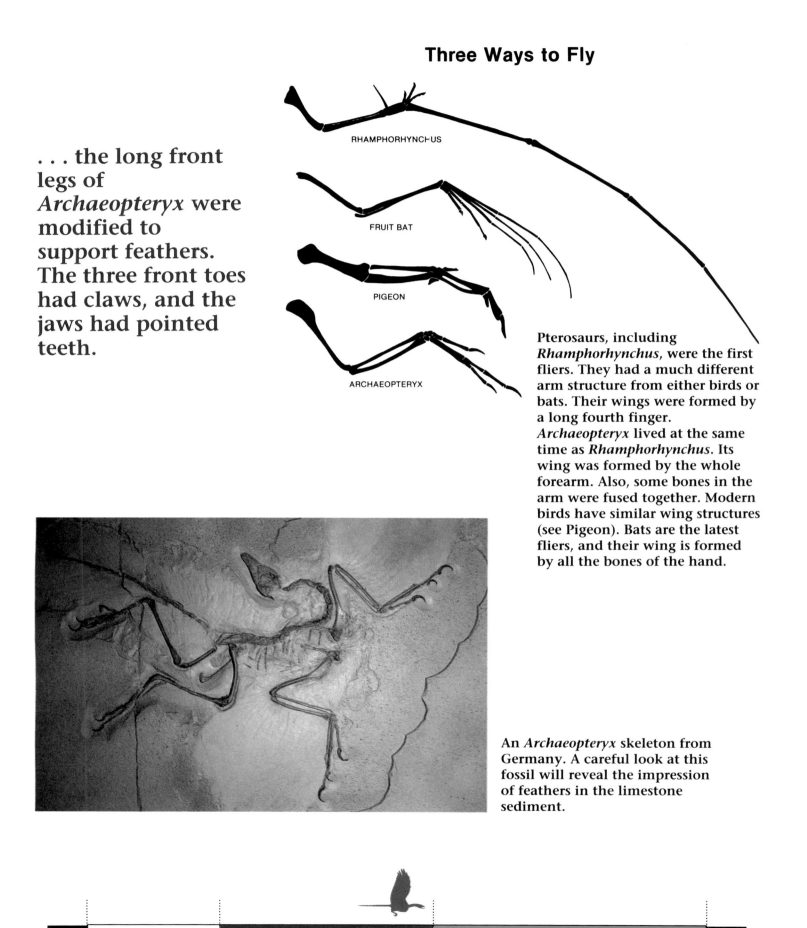

RHAMPHORHYNCHUS

FRUIT BAT

PIGEON

ARCHAEOPTERYX

. . . the long front legs of *Archaeopteryx* were modified to support feathers. The three front toes had claws, and the jaws had pointed teeth.

Pterosaurs, including *Rhamphorhynchus*, were the first fliers. They had a much different arm structure from either birds or bats. Their wings were formed by a long fourth finger. *Archaeopteryx* lived at the same time as *Rhamphorhynchus*. Its wing was formed by the whole forearm. Also, some bones in the arm were fused together. Modern birds have similar wing structures (see Pigeon). Bats are the latest fliers, and their wing is formed by all the bones of the hand.

An *Archaeopteryx* skeleton from Germany. A careful look at this fossil will reveal the impression of feathers in the limestone sediment.

| 245 | TRIASSIC | 208 | JURASSIC | 146 | CRETACEOUS | 65 MILLION YEARS AGO |

CAMARASAURUS

(CAM-er-uh-SORE-us)

Camarasaurus was probably the most common sauropod dinosaur of the Late Jurassic Morrison Formation in North America. This large, 25-ton plant-eater was strong and massive, with powerful legs, a strong neck and tail, and a rounded head. Deep pockets or chambers in the bones of the spine of *Camarasaurus* lightened the skeleton without giving up strength. It is also how the dinosaur got its name, which means "chambered reptile."

The most unusual features of *Camarasaurus* were on its head. The large jaw bones had strong muscles, and the teeth were very large for a sauropod—as big as chisels, with sharp points that chopped the plants that the dinosaur ate. *Camarasaurus* probably fed on plants that were coarse and tough. Its relatives *Apatosaurus* and *Diplodocus,* with their small, weak teeth, probably ate softer, tenderer plants.

With large eyes and nostrils, *Camarasaurus* was alert and active. Like other sauropods, it probably moved in herds. It lived in the arid and semiarid open country of North America.

One *Camarasaurus* pelvis from the Cleveland-Lloyd Quarry in Utah has huge grooves in the bones where an *Allosaurus* tore into the flesh and gouged the bones. *Allosaurus* was *Camarasaurus*'s fiercest enemy, but an adult *Camarasaurus* was so much larger that it was seldom attacked. A complete skeleton of a juvenile *Camarasaurus* was excavated from Dinosaur National Monument in Utah. Such skeletons are rare. Perhaps young sauropods grew to adult size quickly, so there is little chance of finding them in the fossil record.

One interesting twist of fate for *Camarasaurus* was that its head was mistakenly placed on the skeleton of *Apatosaurus* at the Carnegie Museum of Natural History. The mistake was not fixed for 75 years.

Camarasaurus, which moved in herds over dry terrain, was a massive, 25-ton plant-eater. Its most feared predator was *Allosaurus*.

A *Camarasaurus* skull being excavated at Dinosaur National Monument in Utah. The animal's unusually large, chisellike teeth were adapted for chopping up plants.

245 **TRIASSIC** 208 **JURASSIC** 146 **CRETACEOUS** 65
**MILLION
YEARS
AGO**

One interesting twist of fate for *Camarasaurus* was that its head was mistakenly placed on the skeleton of *Apatosaurus* at the Carnegie Museum of Natural History. The mistake was not fixed for 75 years.

Period:
Late Jurassic
Order, Suborder, Family:
Saurischia, Sauropodomorpha, Camarasauridae
Location:
North America
Length:
60 feet (18 meters)

A skeletal drawing of *Camarasaurus*. The name, which means "chambered reptile," refers to pockets, or chambers, in its backbones that lightened the skeleton.

DIPLODOCUS
(dih-PLOH-dah-kus)

Diplodocus got its name because of a feature of its spinal bones. Under each bone of the tail was a piece, called a "chevron," running forward as well as backward, so the tail resembled a "double beam." In other words, *Diplodocus* was built like a giant suspension bridge set between four massive pillars.

Diplodocus is the longest dinosaur known from complete skeletons. It weighed no more than 10 to 12 tons, half of what its larger relatives weighed. This gave *Diplodocus* an advantage in grace and agility when wandering over the landscape. Perhaps it could more easily avoid predators.

The head of *Diplodocus* was small and lightly built. It was long and slender, a little like the shape of a horse's head and about the same size. But, unlike a horse, *Diplodocus* had no teeth in the sides of the jaws. It had long, slender teeth the size and shape of pencils in the front of its mouth.

Clues as to how *Diplodocus* gathered enough food to keep it alive can be found in the skeleton. The neck was long, slender, and flexible. It could reach great heights and places smaller plant-eaters could not. Also, gastroliths or "stomach stones" have been found with some *Diplodocus* skeletons. The gastroliths helped grind food in the stomach to aid digestion.

Early drawings of *Diplodocus* and *Apatosaurus* showed these dinosaurs as slow, plodding reptiles with a sprawling posture, dragging their tails over the ground. Trackways and their skeletons show that this is not correct. *Diplodocus* and its relatives walked with their feet directly beneath their bodies. They also kept their tails off the ground, which balanced their bodies as they walked.

Skeletons of *Diplodocus* discovered by Earl Douglass in eastern Utah at the turn of the century became sensational news. The dinosaur was larger and more impressive than any other in North America, except *Apatosaurus*.

Diplodocus has been on display in more places in the world than any other sauropod. Early in this century, Andrew Carnegie sent casts of the complete skeleton from the Carnegie Museum to the most important museums in Europe.

Period:
Late Jurassic
Order, Suborder, Family:
Saurischia, Sauropodomorpha, Diplodocidae
Location:
North America (United States)
Length:
90 feet (27 meters)

Fossil teeth of *Diplodocus* from Utah's Dinosaur National Monument. The animal's weak jaws could probably only handle soft plants, like ferns and new tree growth. "Stomach stones" may have helped in digestion.

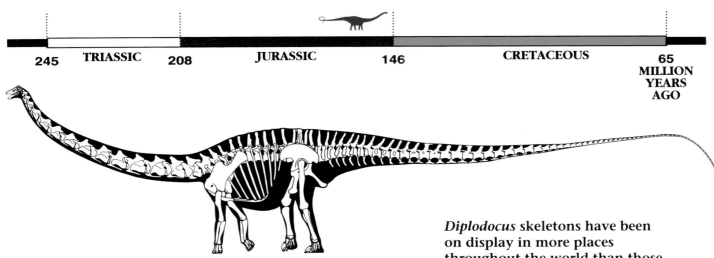

| 245 | TRIASSIC | 208 | JURASSIC | 146 | CRETACEOUS | 65 MILLION YEARS AGO |

Diplodocus is the longest dinosaur known from complete skeletons. It weighed no more than 10 to 12 tons, half of what its larger relatives weighed. This gave *Diplodocus* an advantage in grace and agility when wandering over the landscape. Perhaps it could more easily avoid predators.

Diplodocus skeletons have been on display in more places throughout the world than those of any other sauropod. The one mounted in Pittsburgh's Carnegie Museum is affectionately known as "Dippy."

With a spine built like a suspension bridge over four support columns, the quadrupedal *Diplodocus* walked with its legs directly beneath its body.

STEGOSAURUS

(STEG-oh-SORE-us)

No dinosaur has been the subject of as much controversy as *Stegosaurus,* the armored dinosaur of the Late Jurassic. For example, few ornithischian dinosaurs were armored, and no dinosaur besides *Stegosaurus* and its relatives had huge plates of bone arranged in rows along their backs.

Stegosaurus weighed more than two tons. This plant-eater had few competitors in the Jurassic. *Stegosaurus* preferred food that was near the ground. It was not an agile animal, so it could not compete with other plant-eaters for leaves and twigs higher off the ground.

Stegosaurus was tallest at the hips, which were about ten feet high. The largest bony plates were just behind the hips and added another three feet to its height. This exaggerated its profile, which curved steeply downward both in the front and back. From the side, a predator would have a hard time deciding which end was the head and which was the tail. The head was small, with weak jaws. The front of both jaws was toothless; the dinosaur's beak chopped vegetation.

The front legs of *Stegosaurus* were only half as long as the heavy rear legs, but they were stout and well suited for carrying the weight of the front of the body. The feet were short and stubby, with four blunt toes on the front feet and three toes on the rear.

A double row of flat triangular plates of bone extended from the neck to the tail, which at the tip was armed with two to four pairs of pointed spikes. The plates were several inches thick at the base where they attached to the body, but they were thin and narrow at the tips. Also, smaller knobs and plates in the skin strengthened and protected the flanks and legs.

Stegosaurus has been found only in western North America, but close relatives such as *Kentrosaurus* from Tanzania and *Tuojiangosaurus* from China show a worldwide distribution of this family. Stegosaurs became extinct in North America at the end of the Jurassic, but they survived in other places until late in the Cretaceous Period.

A skeletal drawing of *Stegosaurus.* Pointed spines on its tail gave the rear end of *Stegosaurus* more protection than its front end. The dinosaur's name comes from this armor; *Stegosaurus* means "covered reptile."

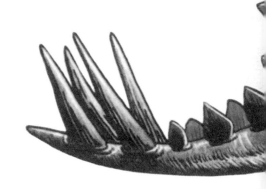

Stegosaurus's bony plates are the subject of scientific argument. Some reconstructions of fossil bones show staggered, rather than paired, plates. And some paleontologists think the plates were arranged single file along the back, rather than in a double row.

. . . few ornithischian dinosaurs were armored, and no dinosaur besides *Stegosaurus* and its relatives had huge plates of bone arranged in rows along their backs.

| 245 | TRIASSIC | 208 | JURASSIC | 146 | CRETACEOUS | 65 MILLION YEARS AGO |

Bones of *Stegosaurus stenops*. The difference in size between front and rear legs shows the dinosaur's bipedal, or two-legged, ancestry. But because of its heavy armor, *Stegosaurus* was forced to walk on all fours.

Period:	Late Jurassic
Order, Suborder, Family:	Ornithischia, Thyreophora, Stegosauridae
Location:	North America
Length:	20–24 feet (6–7 meters)

BRACHIOSAURUS
(BRAK-ee-oh-SORE-us)

Other dinosaurs may be larger (such as *Antarctosaurus, Ultrasauros,* and *Supersaurus*) or longer (such as *Seismosaurus*), but *Brachiosaurus* is the largest sauropod known from nearly complete skeletons. So, for many paleontologists, this is the champion for size. Weighing 80 tons, or about as much as 12 elephants, *Brachiosaurus* was a colossal dinosaur that had to feed constantly.

The front legs were taller than the back legs, and the tail was relatively short. *Brachiosaurus* was heavy in the front and light in the rear. The rib cage was enormous, but because the legs were tall, the belly was so far off the ground that a *Stegosaurus* could walk underneath it.

The long neck and front legs resembled the body of a giraffe, and it is possible that *Brachiosaurus,* like giraffes in Africa today, stood guard in the Jurassic. While it was eating in the high trees, it would have watched for an *Allosaurus* sneaking in for an attack.

Brachiosaurus adults were so enormous that they weighed as much as 40 times more than their principal enemy, *Allosaurus.* Unless an adult was sick or wounded, it had little reason to worry about *Allosaurus,* but a young *Brachiosaurus* had no defense and probably could only run to the center of the herd for protection.

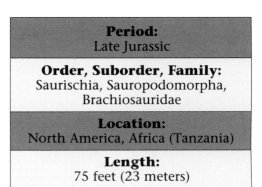

Period:
Late Jurassic
Order, Suborder, Family:
Saurischia, Sauropodomorpha, Brachiosauridae
Location:
North America, Africa (Tanzania)
Length:
75 feet (23 meters)

. . . for many paleontologists, this is the champion for size. Weighing 80 tons, or about as much as 12 elephants, *Brachiosaurus* was a colossal dinosaur that had to feed constantly.

Similarities between *Brachiosaurus* species in North America and Africa suggest the two populations were connected before Pangaea broke apart. If so, the geographic range involved was one of the largest recorded for land animals.

245	TRIASSIC	208	JURASSIC	146	CRETACEOUS	65 MILLION YEARS AGO

CAMPTOSAURUS

(CAMP-toh-SORE-us)

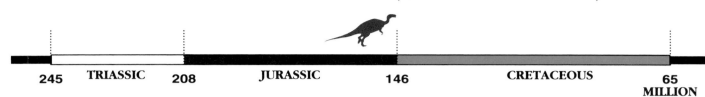

Few plant-eating dinosaurs of the Late Jurassic were small. *Camptosaurus* was an exception; it reached an adult weight of no more than 1,000 pounds. This slender, graceful ornithischian stood around five feet tall at the hips. Some paleontologists compare *Camptosaurus* with a deer in today's forests. *Camptosaurus* browsed on low vegetation. The chisellike teeth on the sides of its jaws were strong and well suited for crushing tough plants. Instead of front teeth for nipping, the front of both jaws was covered by a beak.

The strong, agile rear legs were made for running. It needed to be able to escape from an *Allosaurus* that could easily overpower even the largest *Camptosaurus*. Its front legs were small but strong and were used for slow movement during feeding and grubbing around in the brush. It fed with its short front legs on the ground. The tall hips and rounded curve of the tail gave *Camptosaurus* a curved or bent profile. This is why it got its name, which means "bent reptile."

This early plant-eater is probably closely related to the family Iguanodontidae of the Cretaceous, which included the enormous *Iguanodon* and giant hadrosaurs such as *Parasaurolophus* and *Maiasaura*. Like its descendants, *Camptosaurus* lived in herds, which gave it protection from predators.

About five feet tall at the hips, the plant-eater *Camptosaurus* had an arched profile. Its name means "bent reptile" for that reason.

Period:
Late Jurassic
Order, Suborder, Family:
Ornithischia, Ornithopoda, Camptosauridae
Location:
North America, Europe
Length:
17 feet (5 meters)

Agile and relatively small, *Camptosaurus* lived in herds, which gave it some measure of protection from its natural enemy, *Allosaurus*.

CERATOSAURUS
(seh-RAT-oh-SORE-us)

Ceratosaurus was the greatest rival of *Allosaurus*. *Ceratosaurus* had a fearsome appearance, with a prominent bladelike horn on its snout, knobs in front of its eyes, long daggerlike teeth with jagged edges that sliced through flesh, a huge head with powerful jaws, and a massive body. Its name means "horned lizard."

Ceratosaurus resembled *Allosaurus,* except that its front legs had four fingers rather than three. Each of the fingers and toes had curved claws that could tear flesh from a carcass or bring a victim to the ground with a single blow. Like *Allosaurus,* this meat-eater was an opportunist, probably taking smaller dinosaurs such as *Stegosaurus* or *Camptosaurus* when it could and scavenging on carcasses of the giant sauropods.

Ceratosaurus used its powerful legs and feet for running at high speeds. It could escape danger from other predators or attack its prey by overpowering it with a sudden burst of energy. The huge skull and mighty jaws surely delivered death blows to the victims.

The horns of *Ceratosaurus* may have been for display or for fighting other males of its kind, especially when competing for females. There were not as many ceratosaurs as allosaurs, and they were also not found in as many places. Perhaps *Ceratosaurus* ate only certain animals or needed to stay in one climate.

Period:
Late Jurassic
Order, Suborder, Family:
Saurischia, Theropoda, Ceratosauridae
Location:
North America, Africa (Tanzania)
Length:
20 feet (6 meters)

A *Ceratosaurus* skull. In life, the dinosaur's massive head was frightful, equipped with strong jaws and serrated teeth that could cut flesh like steak knives.

An opportunistic meat-eater and possible scavenger, *Ceratosaurus* had powerful legs built for speed, whether in escaping predators or overtaking prey.

The huge skull and mighty jaws surely delivered death blows to the victims.

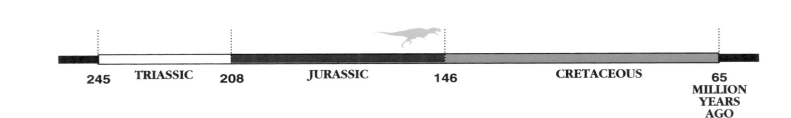

245	TRIASSIC	208	JURASSIC	146	CRETACEOUS	65 MILLION YEARS AGO

CETIOSAURUS
(SEE-tee-oh-SORE-us)

*C*etiosaurus was a heavy sauropod. It looked much like *Diplodocus* and *Apatosaurus,* which may be its descendants. This is the best-known sauropod from England, where dinosaurs were first studied. Only the limbs, pelvis, and greater part of the tail are known. The spinal bones are massive and do not have air pockets that reduce weight, as are found in most other sauropods. Like the legs of other sauropods, the limbs of *Cetiosaurus* resembled pillars; the tail was long and heavy. *Cetiosaurus* may have weighed as much as 30 tons. It was one of the more primitive sauropods.

Cetiosaurus was described in 1841 and was the first sauropod dinosaur described. *Cetiosaurus,* which means "monster reptile" or "whale reptile," was the largest land animal known at the time, much larger than *Iguanodon* and prehistoric reptiles. It was about the same size as the North Atlantic great whales. *Cetiosaurus* was first thought to be a crocodile and was later confused with *Iguanodon,* an Early Cretaceous dinosaur.

Some paleontologists think another British sauropod, *Cetiosauriscus,* belongs to this genus. Bones in the tail of *Cetiosauriscus,* called "chevrons," resemble those of *Diplodocus.* If this is true, there is a close relationship between the families Cetiosauridae and Diplodocidae. This would show a connection among dinosaurs in Morocco, England, and North America during the Jurassic Period.

Period:
Middle Jurassic
Order, Suborder, Family:
Saurischia, Sauropodomorpha, Cetiosauridae
Location:
Europe (England), Africa (Morocco)
Length:
60 feet (18 meters)

Cetiosaurus, one of the more primitive sauropods, was also the largest land animal known to have existed when it was discovered in 1841. Scientists think there may be a close relationship between *Cetiosaurus* and *Diplodocus.*

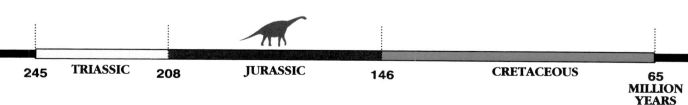

| 245 | TRIASSIC | 208 | JURASSIC | 146 | CRETACEOUS | 65 MILLION YEARS AGO |

COMPSOGNATHUS

(KOMP-sog-NAY-thus)

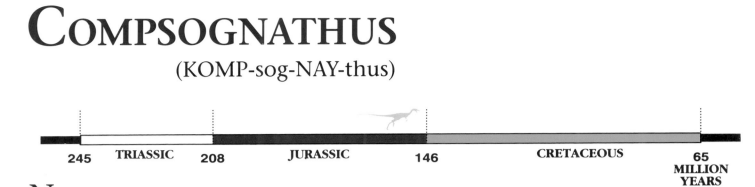

| 245 | TRIASSIC | 208 | JURASSIC | 146 | CRETACEOUS | 65 MILLION YEARS AGO |

Not all meat-eating dinosaurs were the gigantic brutes we imagine. Some were small and delicate. *Compsognathus* was one of the smallest known. German paleontologist Andreas Wagner described the first *Compsognathus* in 1861. Its name means "elegant jaw."

Compsognathus longipes is rare in museum collections. There is only one other specimen besides Wagner's, a skeleton about 50 percent larger discovered in limestone near Canjuers, France. A group of French paleontologists reported on this skeleton in 1972. They thought it belonged to a different species and named this new species *Compsognathus corallestris*. It seemed to have flippers and different skeletal proportions.

Paleontologist John Ostrom showed, however, that the "flipper imprint" in the specimen did not belong to the animal and that the differences between the two specimens' sizes were probably due to differences in age and development. Scientists now think both animals belong to the single species *Compsognathus longipes*.

Compsognathus probably had two-fingered hands. This surprised many paleontologists, because all other known theropods except tyrannosaurids had hands with three or more fingers. Not everyone agrees that Ostrom's interpretation is correct. *Compsognathus* is different enough to be classified as the only member of its family, Compsognathidae. If Ostrom is right, then *Compsognathus* is an even more distant relative of the giant predators than scientists had previously thought.

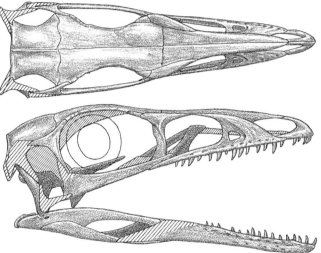

Skull drawings of *Compsognathus*. With its toothy but "elegant" jaws, it may have preyed upon *Archaeopteryx*, the oldest known bird.

A predator that lived near water, *Compsognathus* ran like a bird, chasing lizards and small mammals and attacking its prey with two-fingered hands.

| **Period:** |
| Late Jurassic |
| **Order, Suborder, Family:** |
| Saurischia, Theropoda, Compsognathidae |
| **Location:** |
| Europe (Germany, France) |
| **Length:** |
| 2–3 feet (60 centimeters to 1 meter) |

DACENTRURUS
(DAH-sin-TRUE-rus)

Also known as *Omosaurus, Dacentrurus* was an armored dinosaur related to *Stegosaurus*. It had a set of prominent spikes on its tail, a feature for which this dinosaur was named. *Dacentrurus* means "very spiny tail."

Like *Stegosaurus, Dacentrurus* had plates of bone on its back, but these were more like spikes than triangles. Spikes covered its hips and tail. In this way, *Dacentrurus* was closely related to the African stegosaur *Kentrosaurus. Dacentrurus* is not very well known. It was much smaller than *Stegosaurus*.

As was typical for stegosaurs, *Dacentrurus* was a plant-eater that fed on low-growing vegetation. Its plates and spines were probably its protection from predators. The real function of its armor, and the armor on all stegosaurs, is often a topic of discussion by paleontologists.

Period:
Late Jurassic
Order, Suborder, Family:
Ornithischia, Thyreophora, Stegosauridae
Location:
Europe (England, Western Europe)
Length:
13 feet (4 meters)

Dacentrurus, a relative of *Stegosaurus* that fed on low-growing plants, had bony plates along its back that resembled spikes rather than triangles.

It had a set of prominent spikes on its tail, a feature for which this dinosaur was named. *Dacentrurus* means "very spiny tail."

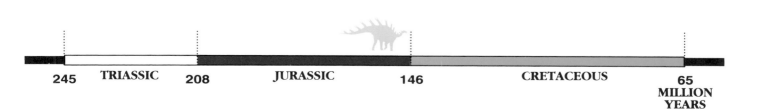

| 245 | TRIASSIC | 208 | JURASSIC | 146 | CRETACEOUS | 65 MILLION YEARS AGO |

DATOUSAURUS

(DAH-too-SORE-us)

A rare Middle Jurassic sauropod from China, *Datousaurus* may have been an animal that traveled by itself and ate the leaves of the tallest plants in its environment. Its neck bones were quite long, giving it a higher reach than *Shunosaurus,* an animal that lived at the same time.

The spines of some of the bones at the base of its neck had a shallow channel for the attachment of muscles and ligaments that controlled the neck's movements. The head was deep and boxlike. The dinosaur looked a bit like the Late Jurassic North American sauropod *Camarasaurus,* to which *Datousaurus* may have been distantly related. Other features of its skeleton may mean that it was related to the diplodocid sauropods.

Only two incomplete skeletons of *Datousaurus bashanensis,* both headless, have been described since 1980. They were found in the Dashanpu Quarry, Sichuan Province, in the People's Republic of China. Since then, more material has been discovered. One skull probably belongs to *Datousaurus,* but scientists will not know until a skeleton is found with the head attached.

Datousaurus was larger than *Shunosaurus,* with bigger, more spoon-shaped teeth. This may mean that *Datousaurus* ate different plants than the other animal.

Period:
Middle Jurassic

Order, Suborder, Family:
Saurischia, Sauropodomorpha, Cetiosauridae

Location:
Asia (People's Republic of China)

Length:
50 feet (15 meters)

. . . Datousaurus may have been an animal that traveled by itself and ate the leaves of the tallest plants in its environment.

Datousaurus bashanensis, a long-necked plant-eater with a deep, boxlike head, had neckbones whose spines allowed attachment of muscles and ligaments for controlling neck movement.

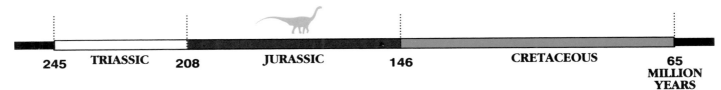

245	TRIASSIC	208	JURASSIC	146	CRETACEOUS	65 MILLION YEARS AGO

DRYOSAURUS

(DRY-oh-SORE-us)

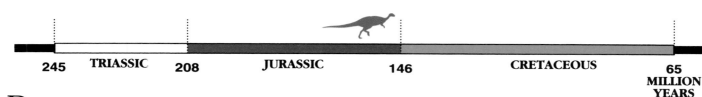

| 245 | **TRIASSIC** | 208 | **JURASSIC** | 146 | **CRETACEOUS** | 65 MILLION YEARS AGO |

Dryosaurus is the most important member of the family Dryosauridae. This is a group of small, plant-eating dinosaurs from the Late Jurassic and Early Cretaceous of North America, eastern Africa, and Europe.

Dryosaurus had long, powerful back legs. Its foot was slim and had three toes. Its arms were strong and had five-fingered hands; the animal probably used its hands to grasp leaves and branches when feeding. Its stiff tail balanced the body while it was standing or running across the countryside.

Dryosaurus had no teeth at the front of its mouth. It used its horny beak to nip ground-level plants and shrubs and low-lying branches. The dinosaur had chewing teeth, located toward the back of the jaws, that ground up leaves and shoots before the animal swallowed them.

Workers found *Dryosaurus* in Late Jurassic rocks in the western United States. Also, a spectacular fossil site of the same age is in Tanzania, Africa. Werner Janensch from Berlin, Germany, collected hundreds of *Dryosaurus* bones from Tanzania in the early part of the 20th century. Many of these fossils were young, even hatchling, dryosaurs. This will tell scientists a great deal about how these dinosaurs grew and developed.

Othniel Charles Marsh coined the name *Dryosaurus* in 1894. It means "oak reptile." *Dryosaurus,* along with its close relative, *Valdosaurus,* was one of the earliest members of a large group of important ornithopods known as iguanodontians. Other members of this group included *Tenontosaurus, Camptosaurus,* and the later iguanodontids and hadrosaurids.

| **Period:** |
| Late Jurassic |
| **Order, Suborder, Family:** |
| Ornithischia, Ornithopoda, Dryosauridae |
| **Location:** |
| North America, Africa (Tanzania) |
| **Length:** |
| 10–13 feet (3–4 meters) |

A skeletal drawing of *Dryosaurus.* The "oak reptile" fed on low-lying plants and shrubs, nipping at them with its horny, toothless beak.

The dinosaur had chewing teeth, located toward the back of the jaws, that ground up leaves and shoots before the animal swallowed them.

An early iguanodontian, *Dryosaurus* had powerful back legs for running and a stiff tail that balanced the animal, at rest or in motion.

EUHELOPUS

(you-HEL-oh-pus)

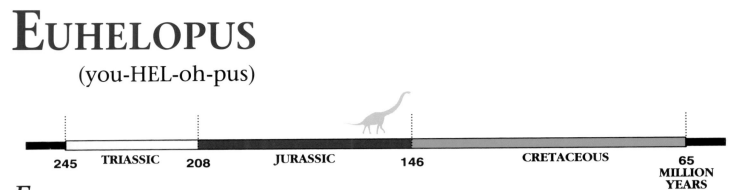

| 245 | TRIASSIC | 208 | JURASSIC | 146 | CRETACEOUS | 65 MILLION YEARS AGO |

*E*uhelopus was a large sauropod from China. It was much like *Camarasaurus* from the Late Jurassic of North America, but it was long and slender, with extra bones in the neck and trunk. The skull of *Euhelopus* had a long snout region, quite unlike the short, blunt head of *Camarasaurus*. Unfortunately, no teeth were found with the skull.

Euhelopus is known from a partial skeleton, including most of the neck and spinal bones. The dinosaur probably weighed around 15 to 20 tons. With its long, slender neck, *Euhelopus* may have looked somewhat like *Brachiosaurus*. It also probably reached high into treetops for food.

Originally named *Helopus,* a name that had been used before, the dinosaur was renamed *Euhelopus* in 1956. The name means "true marsh foot." Because of its unusual combination of features, *Euhelopus* is sometimes classified in a separate family, but the bones of the spine suggest that the dinosaur was closely related to the family Camarasauridae.

Euhelopus zdanskyi. Extra spinal bones made the "true marsh foot" reptile long and slender. It weighed as much as 20 tons.

Euhelopus is known from a partial skeleton, including most of the neck and spinal bones.

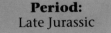

Period:
Late Jurassic
Order, Suborder, Family:
Saurischia, Sauropodomorpha, Camarasauridae
Location:
Asia (People's Republic of China)
Length:
50 feet (15 meters)

HUAYANGOSAURUS

(hoy-YANG-oh-SORE-us)

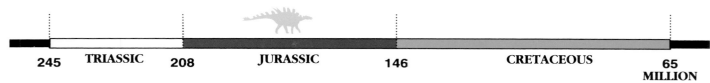

| 245 | **TRIASSIC** | 208 | **JURASSIC** | 146 | **CRETACEOUS** | 65 MILLION YEARS AGO |

*H*uayangosaurus is the most primitive stegosaur. Its skull had a small opening in front of each eye, and there was another small opening in each half of the lower jaw. Both of these openings closed off in later stegosaurs. At the front of its snout, *Huayangosaurus* had 14 teeth (seven on a side). Later stegosaurs lacked teeth in front.

Huayangosaurus had long front limbs—almost as long as the back limbs. Later stegosaurs had forelimbs that were much shorter.

Finally, the armor plates that ran in two rows along the back of *Huayangosaurus* were narrower and much thicker than the plates of its later relatives. All these features are clues to the stegosaurs' place in the dinosaur family tree.

Huayangosaurus lived in a land of lakes, rivers, and lush vegetation. By looking at its teeth, scientists can tell it was a plant-eater. Its spiky, erect armor plates and shoulder spines could have protected the dinosaur from predators, but they could have been for show, perhaps to attract a mate. (They may also have been used to regulate body temperature, although the plates seem too thick to have been very good for this.) Certainly, the animal's tail spikes would have kept its enemies away.

| **Period:** |
| Middle Jurassic |
| **Order, Suborder, Family:** |
| Ornithischia, Thyreophora, Huayangosauridae |
| **Location:** |
| Asia (People's Republic of China) |
| **Length:** |
| 15 feet (4.5 meters) |

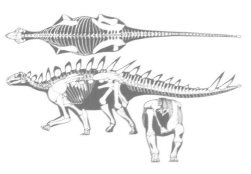

Skeletal drawings of *Huayangosaurus*, the most primitive stegosaur.

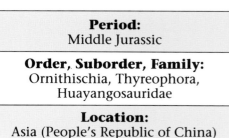

Huayangosaurus taibaii. Recently, a Chinese paleontologist proposed that *Huayangosaurus* and *Tatisaurus,* which have similar jaws, be grouped together in a separate stegosaur subfamily, the Huayangosauridae. More research may show that this subfamily bridges the gap between the stegosaurs and the ankylosaurs.

71

KENTROSAURUS

(KEN-troh-SORE-us)

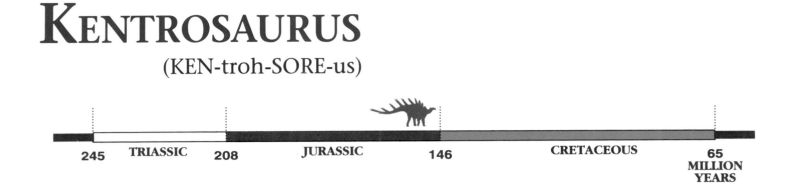

| 245 | TRIASSIC | 208 | JURASSIC | 146 | CRETACEOUS | 65 MILLION YEARS AGO |

*K*entrosaurus was only about half as large as its relative *Stegosaurus*. *Kentrosaurus* had spikes over its back, hips, and tail that discouraged predators—and also gave it its name, which means "prickly reptile." It had an extra pair of spines that stuck out sideways and backward from its shoulders. The tail spines were long and pointed.

This small stegosaur stood up to five feet tall at the hips. Because of its short front legs and short neck, it could only eat the lowest shrubs and plants. Occasionally, *Kentrosaurus* may have leaned backward and stood on its back legs, supported by its tail, to reach for taller plants.

Kentrosaurus lived alongside *Brachiosaurus* in eastern Africa. Both dinosaurs were probably troubled by large, powerful predators, such as *Allosaurus*. The armor of *Kentrosaurus* only partly protected its body; its sides and underbelly were left undefended. *Kentrosaurus* may have had some other ways to protect itself that paleontologists have not yet discovered.

Other stegosaurs of the Late Jurassic, including *Dacentrurus* in Western Europe, *Stegosaurus* in North America, and *Tuojiangosaurus* in China, were closely related. Land connections in the Jurassic may have let stegosaur ancestors travel around the northern continents.

Period:
Late Jurassic
Order, Suborder, Family:
Ornithischia, Thyreophora, Stegosauridae
Location:
Africa (Tanzania)
Length:
10 feet (3 meters)

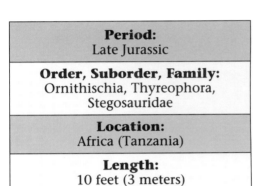

Kentrosaurus. This plant-eating "prickly reptile" probably protected itself from predators like *Allosaurus* by using its spiky armor and spiny tail.

This small stegosaur stood up to five feet tall at the hips. Because of its short front legs and short neck, it could only eat the lowest shrubs and plants.

MAMENCHISAURUS

(mah-MEN-chee-SORE-us)

Period:
Late Jurassic
Order, Suborder, Family:
Saurischia, Sauropodomorpha, Diplodocidae
Location:
Asia (People's Republic of China)
Length:
72 feet (22 meters)

A skeletal reconstruction of *Mamenchisaurus hochuanensis.* The Chinese thought the dinosaur's bones were dragon remains and ground them up for sale in drug stores.

It had extra neck bones, and its spinal bones were longer than usual; this gave *Mamenchisaurus* an almost impossibly long neck of about 33 feet.

Mamenchisaurus is the largest sauropod known from China, and nearly half its length was its neck. The front limbs and skull have not been found, but the rest of the skeleton is complete.

The slender and almost delicate proportions of this sauropod are an architectural wonder. It had extra neck bones, and its spinal bones were longer than usual; this gave *Mamenchisaurus* an almost impossibly long neck of about 33 feet. Add the shoulder height of about 11 feet, and *Mamenchisaurus* could reach about 44 feet off the ground. It would have been tall enough to peek into fourth-story windows.

Dinosaurs have been associated with dragons for a long time. *Mamenchisaurus* was discovered at a collection site for "dragon bones" that were ground up and sold in drug stores. This site in China is called Mamenchi. "Chi" means brook, and "Mamen" was the name of the brook; so *Mamenchisaurus* means "reptile from the brook named Mamen."

This plant-eater weighed around 20 tons and was 72 feet long. The legs and tail of *Mamenchisaurus* were light and slender compared to other sauropods; it was probably agile and graceful when it walked. It may have wandered in herds like other sauropods, always feeding and protecting the young animals from meat-eating dinosaurs waiting to attack.

Replicas of the original skeleton have been shown in many museums around the world. Other dinosaurs from China have been excavated in recent years, and several projects have led to an exchange of information among museums in China, Canada, and the United States.

245	TRIASSIC	208	JURASSIC	146	CRETACEOUS	65 MILLION YEARS AGO

MEGALOSAURUS

(MEG-ah-loh-SORE-us)

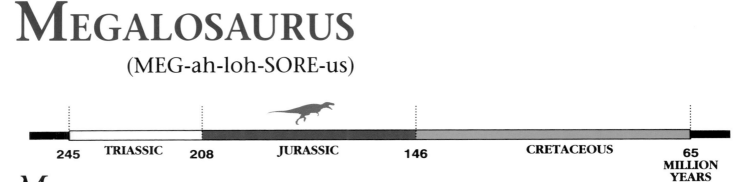

| 245 | TRIASSIC | 208 | JURASSIC | 146 | CRETACEOUS | 65 MILLION YEARS AGO |

Megalosaurus was the first dinosaur to be described as a reptile. The bones were discovered in Oxfordshire, England, in 1818 and fossils included parts of a back leg, hip bones, a shoulder blade, and a lower jaw. William Buckland named it *Megalosaurus,* which means "great lizard." Dinosaurs were not known then, and scientists thought it was a giant lizard.

For many years, scientists identified nearly every scrap of any large theropod fossil in England or Europe as *Megalosaurus.* Even recently, specimens from the United States and Australia were called *Megalosaurus.* For this reason, scientists refer to *Megalosaurus* as the "waste basket" genus. Also, the family Megalosauridae has been the "waste basket" family because scientists included in it dinosaurs as different as *Tyrannosaurus, Allosaurus,* and *Dilophosaurus.* Scientists have sorted out many of the fossils that were incorrectly called *Megalosaurus* or wrongly put in Megalosauridae and placed them in the correct genus and family.

Despite the mistakes, *Megalosaurus* is still an important dinosaur, though it is not well known. Most of the bones of the original specimen are important, particularly the jaw and hip bones. These bones show that it was a large theropod, weighing more than a ton. The large cutting teeth in the robust jaw bone suggest it was a powerful predator.

| **Period:** |
| Middle Jurassic |
| **Order, Suborder, Family:** |
| Saurischia, Theropoda, Megalosauridae |
| **Location:** |
| Europe (England) |
| **Length:** |
| 30 feet (9 meters) |

Fossil bones of *Megalosaurus.* The first skeletal remains, found in 19th-century England, included parts of a leg, hip, and shoulder blade, as well as a jaw with sharp teeth.

William Buckland named it *Megalosaurus,* which means "great lizard." Dinosaurs were not known then, and scientists thought it was a giant lizard.

Like most modern meat-eating animals, *Megalosaurus* probably ate any animal it could capture.

OMEISAURUS
(OH-mee-SORE-us)

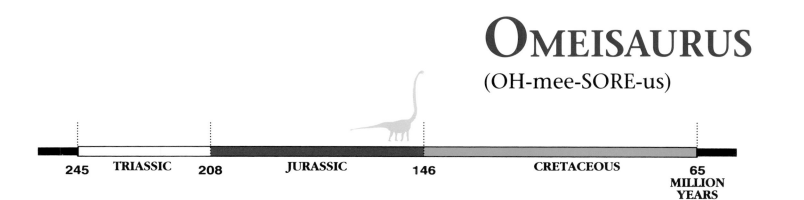

245	TRIASSIC	208	JURASSIC	146	CRETACEOUS	65 MILLION YEARS AGO

Period:
Late Jurassic

Order, Suborder, Family:
Saurischia, Sauropodomorpha,
Cetiosauridae

Location:
Asia (People's Republic of China)

Length:
35–67 feet (10.5–20 meters)

Omeisaurus in a forest setting.
The dinosaur was named after the
sacred Chinese mountain,
Omeishan, near where the first
skeleton was found.

Most of the sauropods known from the Late Jurassic of China had extraordinarily long necks. Not only did their necks have several more bones than the sauropods of North America and Africa, the bones themselves were larger and longer. Herds of *Omeisaurus* roamed through semitropical forests, their necks stretching high into the treetops. Except for the very largest trees, whatever plants the animals did not eat were trampled beneath their enormous feet.

The neck of *Omeisaurus* seems too long when compared with its body; it does not seem balanced. But its spinal bones had thin walls and large holes, making them light. Groups of muscles and long, delicate ribs attached to these bones so that *Omeisaurus* could control its neck.

Omeisaurus had a deep, blunt head and spoon-shaped teeth for eating plants.

Sauropods' huge size probably defended most of them from attack by predatory dinosaurs, but *Omeisaurus* may have had something else to protect it. In the late 1980s, paleontologists working on the Middle Jurassic sauropod *Shunosaurus* discovered that its tail ended in a large bony club, like that of the ankylosaurids. Workers found more tail clubs, of a different shape, in the same locations as *Omeisaurus*. Some Chinese paleontologists think they may belong to *Omeisaurus*.

Omeisaurus had 17 neck bones. The longest neck belonged to *O. tianfuensis;* its neck was about 30 feet long. A 67-foot-long mounted skeleton of *O. tianfuensis* is on display at the Zigong Museum in China. Only the neck of *Mamenchisaurus hochuanensis,* with 19 neck bones, was longer.

75

ORNITHOLESTES
(OR-nith-oh-LEST-ees)

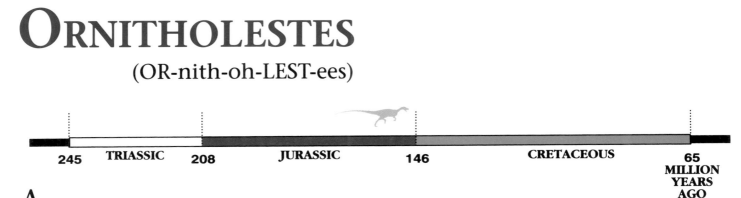

245	TRIASSIC	208	JURASSIC	146	CRETACEOUS	65 MILLION YEARS AGO

A crew from the American Museum of Natural History found the first remains of *Ornitholestes* (or "bird robber") at Bone Cabin Quarry near Medicine Bow, Wyoming, in 1900. It consisted of a skull and partial skeleton.

H. F. Osborn described *Ornitholestes* in 1903. In that paper, he also grouped a partial hand from the same quarry, but not the same animal, in the same genus. The skull is complete, though badly crushed. Jagged cutting teeth line the jaws. The first tooth in the upper jaw is the largest. *Ornitholestes* may have had a small horn over its nose, but scientists are not sure.

The hands are not complete. If the partial hand of the other animal belongs to the same genus, the first finger was short, and the second and third fingers were much longer. *Ornitholestes* may have captured and held its prey with its hands. All the fingers had sharp curved claws. The animal probably weighed about 35 pounds.

For many years, scientists thought *Ornitholestes* and *Coelurus* were the same genus. In 1980, however, John Ostrom showed that they were not the same. Since the discovery of *Ornitholestes* in 1900, workers have not found additional skeletons of this animal. It seems now that *Ornitholestes* was a rare member of its fauna or that it was rarely preserved.

A pair of *Ornitholestes* skeletons. This delicate, 35-pound predator seems to have been either rare or rarely preserved.

Period:
Late Jurassic
Order, Suborder, Family:
Saurischia, Theropoda, Coeluridae
Location:
North America (United States)
Length:
6½ feet (2 meters)

Ornitholestes, or "bird robber," had jagged-edged teeth for cutting flesh and sharp-clawed fingers, perhaps for snaring prey.

OTHNIELIA
(oth-NEE-lee-ah)

Othnielia is a recently named dinosaur. Peter Galton named it in 1977 and described the animal in 1983. *Othnielia* is in honor of Othniel Charles Marsh, the 19th-century American paleontologist. Of Late Jurassic age, *Othnielia* is not a well-known hypsilophodontid dinosaur. The only material we have is three or four skeletons and some pieces of skull and teeth. Unfortunately, no full skull has been found for *Othnielia,* but there is still much information that scientists can piece together.

From the teeth, we know *Othnielia* probably ate leaves, succulent plants, and possibly insects. Its close relatives lived in the Early and Late Cretaceous, among them *Zephyrosaurus, Thescelosaurus,* and *Orodromeus.* These animals also fed on low-growing plants and shrubs.

The limbs of *Othnielia* show that it would have been a good runner. This speed was useful when *Othnielia* needed to escape such predators as *Ornitholestes, Coelurus,* and *Allosaurus.* Speed would also have been needed to catch insects.

During the Late Jurassic, *Othnielia* was part of a group of plant-eating dinosaurs in western North America, including *Camptosaurus, Dryosaurus,* and *Stegosaurus. Othnielia* also lived at the same time as the sauropods *Apatosaurus* and *Diplodocus.*

Period:
Late Jurassic
Order, Suborder, Family:
Ornithischia, Ornithopoda, Hypsilophodontidae
Location:
North America
Length:
10 feet (3 meters)

Unfortunately, no full skull has been found for *Othnielia*, but there is still much information that scientists can piece together. From the teeth, we know *Othnielia* probably ate leaves, succulent plants, and possibly insects.

Othnielia rex. This speedy ornithopod may have used its running ability, not only to flee *Allosaurus* and other predators, but to catch insects.

| 245 | TRIASSIC | 208 | JURASSIC | 146 | CRETACEOUS | 65 MILLION YEARS AGO |

PATAGOSAURUS
(PAH-tag-oh-SORE-us)

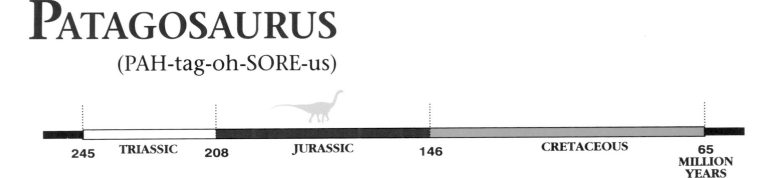

| 245 | TRIASSIC | 208 | JURASSIC | 146 | CRETACEOUS | 65 MILLION YEARS AGO |

Older than the Late Jurassic dinosaurs of North America, *Patagosaurus* is one of only two sauropods known from the Middle Jurassic of South America. The other sauropod is *Volkheimeria*, which is a member of the same family as *Patagosaurus*. A *Patagosaurus* skeleton was found in Patagonia, Argentina, in rocks about 15 million years older than those where the North American sauropods were unearthed.

Patagosaurus is known from a nearly complete skeleton that is lacking the skull. It resembled the English sauropod *Cetiosaurus*, but its hips and spinal bones were different. *Patagosaurus* was a medium-size sauropod that weighed about 15 tons. In some ways it looked like the North American sauropod *Haplocanthosaurus*, but it was more primitive. *Patagosaurus* and its relatives may be the direct ancestors of some of the Late Jurassic sauropods. There may have been land connections that allowed the sauropods to travel between South America, Western Europe, and Africa. Members of the same family have been found in these different places.

Patagosaurus ate plants, and its enemies were the theropod dinosaurs. *Piatnitzkysaurus*, a theropod found at the same time and in the same location as *Patagosaurus* and *Volkheimeria*, was as large as *Allosaurus* and a menace to most plant-eaters.

Period:
Middle Jurassic
Order, Suborder, Family:
Saurischia, Sauropodomorpha, Cetiosauridae
Location:
South America (Argentina)
Length:
50 feet (15 meters)

Patagosaurus fariasi. A 15-ton plant-eater, *Patagosaurus* is a rare Middle Jurassic sauropod from Argentina considered more primitive than its North American relative, *Haplocanthosaurus.*

SEISMOSAURUS

(SIZE-moh-SORE-us)

Nearly half as long as a football field, *Seismosaurus* is the newest supergiant dinosaur to be discovered. The skeleton is currently being excavated in central New Mexico. It was given its name because of its great size—"earth shaker reptile."

This giant sauropod reached an estimated length of 130 to 170 feet. If this is correct, it is a record. The skeleton is mostly joined and consists of the front half of the tail, the pelvis and sacrum, and the spinal bones in the rib-bearing region. In future excavations, paleontologists hope to recover the front legs, neck, and skull.

Scientists have not yet decided what relationship *Seismosaurus* had to the supergiant *Supersaurus* from the Dry Mesa Quarry in Colorado, which looked like *Diplodocus*.

The same bones from each animal have not been found so scientists cannot yet make a bone-to-bone comparison. However, none of the many dinosaur bones from the Dry Mesa Quarry are similar to those of *Seismosaurus*.

Seismosaurus is a member of the family Diplodocidae. Like other diplodocids, *Seismosaurus* probably had a long slender neck, large bulky body, short front legs, tall rear legs, and a long heavy tail. "Stomach stones," or gastroliths, have been found with the skeleton. It is one of the few joined sauropod skeletons that had gastroliths in place when it was excavated.

Period:
Late Jurassic
Order, Suborder, Family:
Saurischia, Sauropodomorpha, Diplodocidae
Location:
North America (United States)
Length:
130–170 feet (40–52 meters)

It was given its name because of its great size—"earth shaker reptile." This giant sauropod reached an estimated length of 130 to 170 feet. If this is correct, it is a record.

Seismosaurus halli. The gigantic sauropod is related to *Diplodocus*, but is much longer and heavier. Scientists are unsure about the relationship of the two diplodocids.

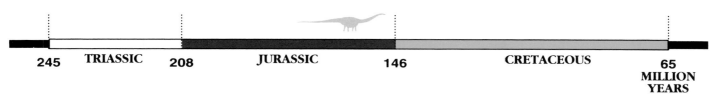

| 245 | **TRIASSIC** | 208 | **JURASSIC** | 146 | **CRETACEOUS** | 65 **MILLION YEARS AGO** |

79

SHUNOSAURUS

(SHOE-noh-SORE-us)

Period:
Middle Jurassic
Order, Suborder, Family:
Saurischia, Sauropodomorpha, Cetiosauridae
Location:
Asia (People's Republic of China)
Length:
40 feet (12 meters)

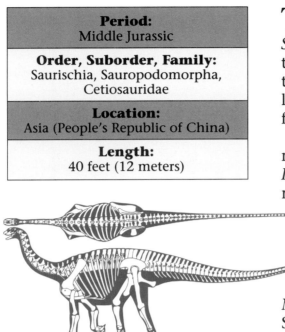

More than 20 nearly complete *Shunosaurus* skeletons have been excavated since the animal's discovery in 1979, including five good skulls. *Shunosaurus* is now one of the best-known sauropods.

The dominant plant-eater of Middle Jurassic China, *Shunosaurus* may have roamed in large herds. With its long, thick neck, it could feed on plants and leaves high above the ground that most other dinosaurs could not reach. Its large body, weighing six to ten tons, made it almost safe from attack by predators, such as the theropod *Gasosaurus*.

Shunosaurus seems to have been a cetiosaurid. It is a relative of the British *Cetiosaurus*, the Argentine *Patagosaurus*, and the Australian *Rhoetosaurus*. Although the neck of *Shunosaurus* was long, it was not as long as the necks of the Late Jurassic sauropods of China and North America. But the skull of *Shunosaurus* shows that it may be near the ancestry of some of the long-necked Chinese sauropods, such as *Euhelopus*, *Omeisaurus* (which may also have had a tail club), and *Mamenchisaurus*. It was also closely related to *Datousaurus*. Since it had many slender teeth and a few "double-beam" chevron bones in its tail, *Shunosaurus* may also have been related to the later North American diplodocids, such as *Diplodocus*, *Apatosaurus,* and *Barosaurus*.

Dinosaur-bearing Middle Jurassic rocks are rare, so there is a gap in our understanding of dinosaur evolution. To date, the remains of more than 100 dinosaurs have been unearthed from the Dashanpu Quarry in Sichuan Province, where *Shunosaurus* remains have been found.

Shunosaurus lii. The dinosaur's tail ended in a large bony club with four short spikes. The tail was probably used as a weapon to scare predators.

245	**TRIASSIC**	208	**JURASSIC**	146	**CRETACEOUS**	65 **MILLION YEARS AGO**

SUPERSAURUS

(SOO-per-SORE-us)

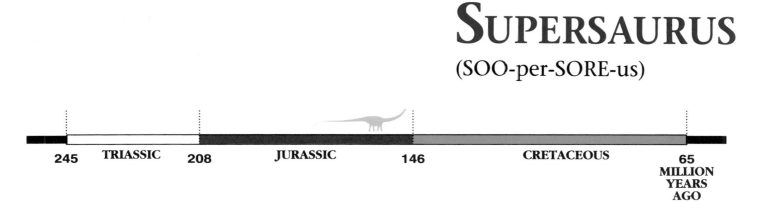

| 245 | TRIASSIC | 208 | JURASSIC | 146 | CRETACEOUS | 65 MILLION YEARS AGO |

Supersaurus is known from a single shoulder bone that is nearly eight feet long. Only the shoulder bone of *Ultrasauros* is as long or longer.

Period:
Late Jurassic

Order, Suborder, Family:
Saurischia, Sauropodomorpha, Diplodocidae

Location:
North America (United States)

Length:
100 feet (30 meters)

Supersaurus deserves its name, "super reptile." A relative of *Apatosaurus* and *Diplodocus*, this enormous plant-eater was rare in the Late Jurassic. *Supersaurus* is known from a single shoulder bone that is nearly eight feet long. Only the shoulder bone of *Ultrasauros* is as long or longer. If restorations of *Supersaurus* are correct, this sauropod was at least 100 feet in length and stood more than 18 feet tall at the shoulders.

The size of *Supersaurus* is difficult to estimate because only the two shoulder bones definitely belong to this genus; the other bones are not positively from *Supersaurus,* so they cannot be used to figure sizes.

The feeding habits of *Supersaurus* were probably like those of other sauropods. It ate leaves and shoots from treetops and ferns from the ground, and it fed almost constantly. Adults must have eaten hundreds of pounds each day. "Stomach stones" may have helped grind the plants and leaves in the stomach.

Supersaurus is a puzzle that can only be solved by more excavations at the Dry Mesa Quarry. More than a dozen dinosaurs are known from the site, each having several hundred bones per animal. With such a mix-up of bones, many years of excavation and careful study will be necessary to find out the identity of this colossal dinosaur.

Stretching its neck up into the trees for soft young shoots, *Supersaurus* could reach 50 feet above the ground, or about the height of a fifth-story window.

TUOJIANGOSAURUS

(toh-HWANG-oh-SORE-us)

Stegosaurs such as *Tuojiangosaurus* may not have had bony spikes and plates when they hatched. The armor may have developed slowly, growing fastest as the animal reached maturity.

The slow-moving, peaceful life of the plant-eating *Tuojiangosaurus* was sometimes interrupted by battles with predators such as *Yangchuanosaurus* and *Szechuanosaurus*. It would also battle with another male *Tuojiangosaurus* for females.

Stegosaurs such as *Tuojiangosaurus* may not have had bony spikes and plates when they hatched. The armor may have developed slowly, growing fastest as the animal reached maturity. Different genera and species probably had different arrangements of plates and spines. This may have helped animals from the same species recognize each other.

The fossils of two *Tuojiangosaurus* were found in the mid-1970s from the Wujiaba Quarry in the Shangshaximiao Formation near the Tuojiang, a river in the Sichuan Basin. One animal, the first almost complete stegosaur skeleton found in China, was mounted there in early 1977. The other specimen was only a set of five spinal bones. The skeleton of a third animal was discovered later, but has not yet been prepared.

Plant-eating dinosaurs that lived at the same time as *Tuojiangosaurus* included its cousin stegosaurs *Chialingosaurus* and *Chungkingosaurus*, the sauropods *Omeisaurus* and *Mamenchisaurus*, and the small ornithopods *Gongbusaurus* and *Yandusaurus*.

Period:
Late Jurassic
Order, Suborder, Family:
Ornithischia, Thyreophora, Stegosauridae
Location:
Asia (People's Republic of China)
Length:
20 feet (6 meters)

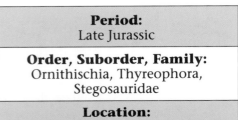

Tuojiangosaurus multispinus probably had 17 pairs of thick, narrow, pointed plates. The last two pairs were thin, cone-shaped spines at the end of its tail. There was also a large, platelike spine above each shoulder.

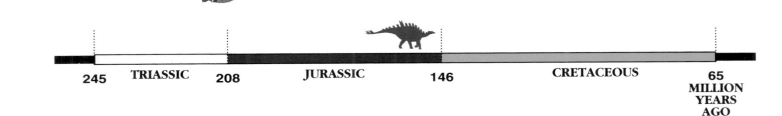

245	TRIASSIC	208	JURASSIC	146	CRETACEOUS	65 MILLION YEARS AGO

ULTRASAUROS
(UL-tra-SORE-us)

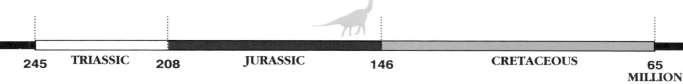

245	**TRIASSIC** 208	**JURASSIC** 146	**CRETACEOUS** 65 MILLION YEARS AGO

Ultrasauros was enormous, which is how it got its name, "ultra reptile." It could be the largest dinosaur known. *Ultrasauros* could have peered into fifth- and sixth-story windows. Walking in a line, only two or three could have fit in a city block, and they were so large that they would have had difficulty turning around except in an intersection.

Like *Supersaurus,* which also came from the Dry Mesa Quarry in southern Colorado, *Ultrasauros* is known only from a few bones, including one from the rib region. Other bones from the quarry may also belong to *Ultrasauros,* but scientists are not yet positive. The other bones include a shoulder bone, a neck bone, and several bones from the tail. The shoulder bone is about nine feet long.

The tall front legs of *Ultrasauros* reached a height at the shoulder of perhaps 25 feet, and the long, massive neck could have reached nearly 60 feet above the ground. The length of *Ultrasauros* was probably close to 90 feet.

Ultrasauros was a plant-eater, like all sauropods. It ate leaves and needles of tree ferns and conifers and any other plant it could reach. At more than 80 tons, *Ultrasauros* fed almost constantly, always on the move in search for food. Herds of *Ultrasauros* must have devastated every forest they entered.

Excavating an *Ultrasauros* shoulder bone. *Ultrasauros* had proportions like a giraffe: The front legs were longer than the rear, and the greatest mass of the body was in front.

Period:	Late Jurassic
Order, Suborder, Family:	Saurischia, Sauropodomorpha, Brachiosauridae
Location:	North America (United States)
Length:	90 feet (27 meters)

Ultrasaurs (right) dominate this Late Jurassic scene. Though *Allosaurus* and others preyed upon it, *Ultrasauros,* at 80 tons, had little to fear.

YANDUSAURUS

(YAN-doo-SORE-us)

Yandusaurus, meaning "reptile from Yandu," is very important in the history of hypsilophodontid dinosaurs. In the late 1970s and early 1980s, workers found it in the Sichuan Province in the People's Republic of China. Paleontologists are studying several skeletons and skulls of *Yandusaurus*. It is the earliest known hypsilophodontid and was found in Middle Jurassic rocks.

This small animal had a short snout, and the back of the skull was high and wide. The spaces for the eyes were large, which might mean that *Yandusaurus* had good eyesight. It had many small triangular teeth in its jaws. The teeth had ridges on their surfaces, similar to the teeth of *Thescelosaurus*. The teeth and jaws show that *Yandusaurus* was a successful plant-eater. It may have also eaten slow-moving insects.

Its neck was long and the body somewhat thinly built. The front limbs were strong with large, clawed hands. Scientists do not know whether the hands were able to grasp.

The back limbs were very long and athletic, much like *Yandusaurus's* close relative, *Orodromeus*. Its legs were built for running fast. This must have been how *Yandusaurus* was able to escape its fierce predators, such as *Szechuanosaurus* and *Yangchuanosaurus*. The feet of *Yandusaurus* also had claws.

The base of the tail is present in only one specimen. The long tail may have acted as a balance for the body while the animal was running.

Period:
Middle Jurassic
Order, Suborder, Family:
Ornithischia, Ornithopoda, Hypsilophodontidae
Location:
Asia (People's Republic of China)
Length:
6½ feet (2 meters)

A skeleton of *Yandusaurus dashanpensis*. Note its graceful hind limbs, built for running fast.

The teeth and jaws show that *Yandusaurus* was a successful plant-eater. It may have also eaten slow-moving insects.

The "reptile from Yandu," *Yandusaurus* is the earliest known hysilophodontid. Speed and good eyesight were its greatest assets.

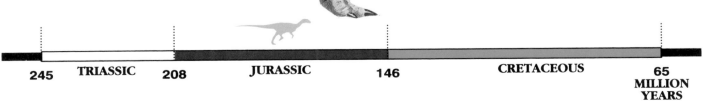

| 245 | TRIASSIC | 208 | JURASSIC | 146 | CRETACEOUS | 65 MILLION YEARS AGO |

YANGCHUANOSAURUS

(yang-choo-AHN-oh-SORE-us)

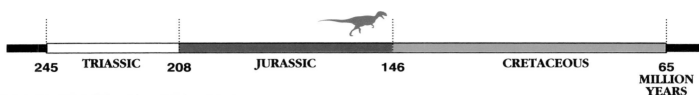

| 245 | TRIASSIC | 208 | JURASSIC | 146 | CRETACEOUS | 65 MILLION YEARS AGO |

Period:	Late Jurassic
Order, Suborder, Family:	Saurischia, Theropoda, Allosauridae
Location:	Asia (People's Republic of China)
Length:	Estimated 30 feet (9 meters)

Until 1976, we only knew the predatory dinosaurs of China from teeth and pieces of fossil bones. Then a construction worker in Yangchuan Province found the first nearly complete skeleton of an *Allosaurus*-size meat-eater while digging the foundation for the Shangyou dam.

When scientists studied the skeleton, only its small forelimbs and the back half of its tail were missing. In 1978, a group of Chinese paleontologists named the dinosaur *Yangchuanosaurus shangyouensis*. Museum workers prepared and mounted its skeleton at the Chongqing Municipal Museum.

Since the first find, workers have discovered a much larger *Yangchuanosaurus* skeleton nearby. It has been described as a second species, *Y. magnus*. With a skull more than three feet long, it is the largest Late Jurassic theropod known from China. Many paleontologists think it is a fully grown *Y. shangyouensis,* because it was in rocks of the same age. It differs from the original species because it is larger and has more fully developed bones.

Like many predatory dinosaurs, *Yangchuanosaurus* had a low crest that ran along the front of its snout, from the nose to slightly in front of the eyes. In life, the crest probably had a horny covering. It may have been brightly colored to attract a mate. *Yangchuanosaurus* almost certainly hunted the plant-eaters of its day, including the stegosaur *Chungkingosaurus.*

Yangchuanosaurus, a meat-eater, bore a nasal crest that may have distinguished males from females.

A *Yangchuanosaurus* skeleton. Scientists are unsure whether a large specimen found after the 1976 discovery is a new species or just the adult version of the earlier find.

85

CHAPTER 5
DINOSAURS BUILD THEIR EMPIRE
Early Cretaceous Period

When the Cretaceous Period began 146 million years ago, the huge land mass known as Pangaea had already started to separate. Laurasia (present-day North America, Europe, and Asia) was almost completely separated from Gondwanaland (present-day South America, Africa, India, Antarctica, and Australia). The latter two continents were separated from Africa and moved southeast across the Indian Ocean. India headed northeast. Africa was separated from South America, which became an island continent. Only Eurasia stayed attached to North America during the Early Cretaceous.

The world of that time was warm. There were wet and dry seasons rather than summer and winter. Most areas of the world were covered by tropical and semitropical forests.

Surprisingly modern-looking frogs, salamanders, turtles, and crocodiles lived in the rivers and lakes. Snakes had only begun to evolve, but there were many lizards, along with primitive furry mammals. All these animals provided food for small predatory dinosaurs.

As Laurasia and Gondwanaland broke into smaller continents, dinosaurs on the separated continents evolved differently. England and Belgium contain the best-studied Early Cretaceous rocks. The most famous plant-eater of this time was *Iguanodon*.

Since North America was still joined to Europe (via Greenland), it had dinosaurs quite similar to those of

Early Cretaceous map.

Deinonychus, a small, agile meat-eater, hunted the ornithopod *Tenontosaurus*, perhaps in packs.

86

Europe. *Iguanodon* is known from Early Cretaceous rocks of the western United States. *Hypsilophodon* was also present. Huge sauropods were in decline; the stegosaurs were almost completely gone.

Large ornithopods were not common in the Early Cretaceous of eastern Asia. The iguanodontid *Probactrosaurus* is thought to be close to the ancestry of the duckbilled dinosaurs. Of roughly the same age was a massive, large-nosed iguanodontid, *Iguanodon orientalis*. One common plant-eater in Asia was *Psittacosaurus;* it was a small, two-legged dinosaur. Armored plant-eaters were also common. One of the last stegosaurs, *Wuerhosaurus,* was a 25-foot-long plant-eater.

Psittacosaurus mongoliensis, one of the smallest and most primitive ceratopsians.

Dinosaurs from the Early Cretaceous of Gondwanaland are less well known than those of Laurasia and show some intriguing and unexpected differences. Sauropods remained prominent in the southern hemisphere despite their drastic decline in the northern hemisphere. Large ornithopods were more scarce than their northern relatives; small ornithopods were common. Theropods in the south were often from families different from those in the north.

Dinosaurs from northern Africa had sails on their backs, including *Ouranosaurus* (an iguanodontid), *Rebbachisaurus* (a sauropod), and *Spinosaurus* (a theropod). With their wide surface area, these sails would have helped the animals keep cool, particularly if they stood in the shade with breezes blowing by.

In the Early Cretaceous, many new dinosaurs appeared. During this period, dinosaurs not only maintained but also expanded their "empire."

Baryonyx, an Early Cretaceous theropod. Some paleontologists think it was primarily a fish-eater.

A carnosaur stalks *Iguanodon* in a Cretaceous forest.

87

BARYONYX

(BEAR-ee-ON-icks)

Baryonyx was found in 1983 by an amateur fossil collector in Surrey, England. He discovered a large claw that was nearly a foot long, and the animal was named for this fossil. *Baryonyx* means "heavy claw."

Paleontologists from the British Museum of Natural History went to the clay pit where the first fossil was found, and they discovered almost all of the skeleton. At first, scientists thought *Baryonyx* was unique and should be placed in its own family, the Baryonychidae. Now it seems that it is related to *Spinosaurus,* and so it is in the family Spinosauridae. *Baryonyx* is the most nearly complete theropod skeleton ever found in England.

The body and back legs of *Baryonyx* were much like those of other predatory dinosaurs, but not much else was the same. Unlike most meat-eaters, its arms were long and heavily built. Also, the claws of the hands, especially the claw of the inside finger, were very heavily built. The length and robustness of the arm may mean that *Baryonyx* walked on all four limbs some of the time. If this is true, *Baryonyx* is the only known theropod that did so.

The skull is more surprising than the arms. Most theropods had skulls that were a little longer than they were high, and they usually had 16 teeth in each side of the jaw. *Baryonyx,* however, had a very long, low skull. The lower jaw was slender and had 32 teeth. While most theropods had a U- or V-shaped snout (when viewed from above or below), the snout of *Baryonyx* was spoon-shaped. The shape of the snout and the jagged edges of the teeth were more like those of a fish-eating crocodile than of most dinosaurs.

Another strange feature of *Baryonyx* is that the nasal openings were behind the snout, rather than near its tip as in other theropods. *Baryonyx* also had a long neck, unlike most other large, meat-eating dinosaurs. From these features, scientists think *Baryonyx* probably was a fish-eater. It may have wandered along river banks, stretching its long neck out over the water and using its large claws to catch fish that swam by.

Baryonyx may have been quadrupedal. The dinosaur's snout and teeth somewhat resemble those of today's crocodile.

Period:
Early Cretaceous
Order, Suborder, Family:
Saurischia, Theropoda, Spinosauridae
Location:
Europe (England)
Length:
31 feet (9.4 meters)

A skeletal drawing of *Baryonyx walkeri*. The most nearly complete theropod skeleton from England, *Baryonyx* is distinguished by its long, heavily built arms and the strong claws on its hands.

As a predator, *Baryonyx* may have eaten fish rather than flesh, using the "heavy claws" for which it was named to hook its dinner.

The length and robustness of the arm may mean that *Baryonyx* walked on all four limbs some of the time. If this is true, *Baryonyx* is the only known theropod that did so.

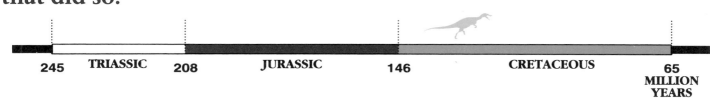

| 245 | TRIASSIC | 208 | JURASSIC | 146 | CRETACEOUS | 65 MILLION YEARS AGO |

CARNOTAURUS

(CAR-noh-TORE-us)

This recently described genus was an unusual theropod. *Carnotaurus* (its name means "meat bull") is known from a single, nearly complete skeleton that had skin impressions over much of the skull and body. It was so complete because it was preserved in a lump of rock that protected the bones and skin impressions. It was found in Early Cretaceous rocks in the Patagonia region of Argentina.

The dinosaur's skull had stout horns above very small eye sockets. Its bladelike, jagged teeth were much like those of other theropods. The arms of *Carnotaurus* were short for its body size, but, unlike *Tyrannosaurus* with its slender arm bones, the arms of *Carnotaurus* were stout. Its arms were so short that it almost looked as if the hands were attached to the upper arm bone. The bones in the back and tail were also unusual; they look a little like the wings of a fantasy spaceship.

The most exciting feature of this animal is its skin. It is more common to find skin impressions of the duckbilled dinosaurs; theropod skin impressions are rare. The skin impressions of *Carnotaurus* show that its skin was made of many low, disklike scales with larger, half-cone-shaped scales in rows along its sides. Like the skin of all known dinosaurs, these scales did not overlap like those on some lizards and snakes.

Carnotaurus and other South American Cretaceous dinosaurs were much different from related animals in other areas of the world, even North America. This group of animals supports the theory that during the Cretaceous Period South America was isolated from the rest of the world, so the animals evolved differently.

The unusual and exciting Cretaceous dinosaurs of South America are just beginning to be worked on and will reveal a great deal of information in the near future.

Period:
Early Cretaceous
Order, Suborder, Family:
Saurischia, Theropoda, Abelisauridae
Location:
South America (Argentina)
Size:
21 feet (6.5 meters)

The skin impressions of *Carnotaurus* show that its skin was made of many low, disklike scales with larger, half-cone-shaped scales in rows along its sides.

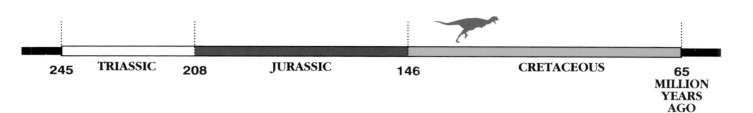

| 245 | TRIASSIC | 208 | JURASSIC | 146 | CRETACEOUS | 65 MILLION YEARS AGO |

A skeleton of *Carnotaurus sastrei* in Buenos Aires, Argentina. A skeleton unearthed in the country's Patagonia region showed skin impressions—a rare find with theropod fossils.

Carnotaurus sastrei. Known for the pattern of scales on its skin, the "meat bull" is very different from Early Cretaceous relatives found outside South America.

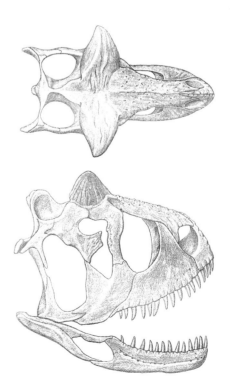

Skull drawings of *Carnotaurus*, top and side view. The skull had small eye sockets, horns above the sockets, and jagged cutting teeth.

IGUANODON

(ih-GWAN-oh-DON)

When the term *Dinosauria* was coined in 1841, it included only three animals: *Megalosaurus*, *Hylaeosaurus*, and *Iguanodon*. *Iguanodon* was first known only from several teeth, which were found by Mary Ann Mantell in England. The animal was later described and named by her husband, Dr. Gideon Mantell, in 1825. It got its name, "iguana tooth," because the tooth looked like that of an iguana. In the 1870s, many complete skeletons of adults and younger animals were found. From all this material, we know a great deal about the behavior, anatomy, and evolution of this important dinosaur.

Iguanodon was a large ornithopod. It walked on its stocky back legs, but the largest animals often walked on all four legs. The front legs were strongly built and the hands specially constructed. Like the toes of the back feet, the fingers had blunt hooves, much like those of today's cows and horses. The outer finger was small and could have been used to grasp, like our thumb. However, the large *Iguanodon* thumb was sharp and like a spike. Scientists are not sure what this spiked thumb was for. It may have been used during sparring matches, perhaps between rival males competing for females, territory, or resources, or it may have been used as a weapon.

As in other ornithopods, the front half of the body was balanced by the long, rigid tail, reinforced by bony tendons. This balance was needed when the animal walked on its back legs. Also, the bony tendons kept the tail from dragging on the ground.

Iguanodon was a plant-eater. The front end of the snout was blunt and had no teeth. Instead, it had a horny covering and was used to crop leaves, shoots, and small branches. Once *Iguanodon* had gotten food into its mouth, the many broad teeth in the back of the jaw tore apart the plants. Muscular cheeks kept food from falling out of the sides of its mouth.

Groups of nearly complete *Iguanodon* skeletons jumbled into piles have been found in Belgium and Germany. These groups of animals may be the remains of small herds of *Iguanodon* caught by a flash flood or while trying to cross a river.

Although *Iguanodon* has been found in many places in Europe, the newest member of the genus was discovered in South Dakota. This species is probably the most primitive.

Iguanodon is the best-known member of the family Iguanodontidae. Another member is *Ouranosaurus,* from Niger in western Africa.

Period:
Early Cretaceous
Order, Suborder, Family:
Ornithischia, Ornithopoda, Iguanodontidae
Location:
Europe (England, France, Germany, Belgium, Spain), North America (United States)
Length:
33 feet (10 meters)

An *Iguanodon* with its young. This famous dinosaur, named for its iguanalike tooth in 1825, has been found throughout Western Europe and most recently in South Dakota.

| 245 | TRIASSIC | 208 | JURASSIC | 146 | CRETACEOUS | 65 MILLION YEARS AGO |

Skeletal drawings of *Iguanodon*. Though it walked on all fours, it could also walk on its hind legs. For this, the rigid, tendon-reinforced tail provided extra balance.

The plant-eater *Iguanodon* had fingers and toes with blunt hooves, like cows and horses today, but it also had a spiked thumb. It lacked teeth in front, but had a horny covering on its snout for cropping vegetation.

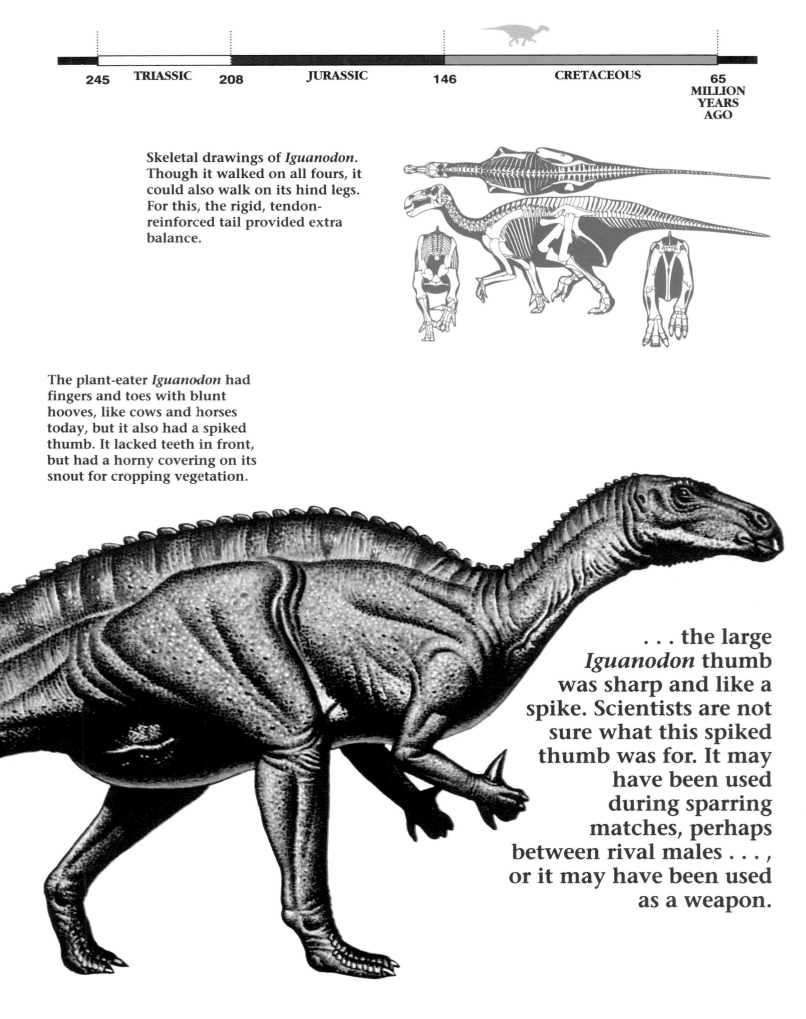

. . . the large *Iguanodon* thumb was sharp and like a spike. Scientists are not sure what this spiked thumb was for. It may have been used during sparring matches, perhaps between rival males . . . , or it may have been used as a weapon.

TENONTOSAURUS
(teh-NON-toh-SORE-us)

Tenontosaurus tilletti was a medium-size ornithopod dinosaur from Montana and Wyoming. Barnum Brown of the American Museum of Natural History discovered the first *Tenontosaurus* skeleton in Montana in 1903. Since then, several dozen partial to complete skeletons have been found, as well as parts of many more.

Some of the skeletons of the younger animals were found jumbled together in groups—three in one group and four in another. The group of four was found with an adult *Tenontosaurus*. These *Tenontosaurus* juveniles may have gathered together or stayed in family groups after they hatched, possibly for protection from predators.

Tenontosaurus had an extremely long, deep tail that was stiffened by "ossified tendons"—tendons that had turned to bone. About two-thirds of the entire length of *Tenontosaurus* was tail. Ossified tendons were also present along its back and over its hips. Even though it probably did most of its walking and running on its back legs, *Tenontosaurus* also had very strong front legs with short, wide front feet. It probably used them to walk on all four limbs to browse on low vegetation.

Tenontosaurus lived at a time when the weather was quite warm and seasonal, with some rain. Common plants included cycads, ferns, and conifers; flowering plants had just begun to evolve. While browsing for food, *Tenontosaurus* probably had to keep watch for the small, fast meat-eating dinosaur *Deinonychus*. We know that *Deinonychus* preyed on *Tenontosaurus* because broken *Deinonychus* teeth have been found with some *Tenontosaurus* skeletons. Some paleontologists believe that *Deinonychus* may have hunted *Tenontosaurus* in packs.

Other animals that lived alongside *Tenontosaurus* included the ankylosaur *Sauropelta*, the coelurid *Microvenator*, and the hypsilophodontid *Zephyrosaurus*. *Tenontosaurus* seems to have been most closely related to the iguanodontids *Dryosaurus*, *Iguanodon*, and *Camptosaurus*.

A *Tenontosaurus* adult and juvenile, feeding on ferns typical of plant life during the Early Cretaceous. *Tenontosaurus's* fiercest foe was *Deinonychus*, which may have hunted the plant-eater in packs.

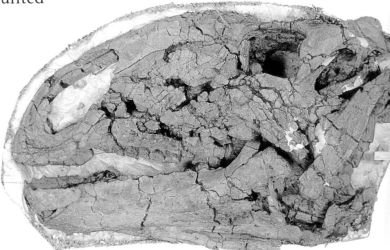

A *Tenontosaurus* skull. The dinosaur lacked teeth in the front part of its mouth, and the horny beak probably bit off plants. Rows of strong, tightly fitted teeth ground up even tough vegetation.

Even though it probably did most of its walking and running on its back legs, *Tenontosaurus* also had very strong front legs with short, wide front feet. It probably used them to walk on all four limbs to browse on low vegetation.

Period:
Early Cretaceous
Order, Suborder, Family:
Ornithischia, Ornithopoda, Hypsilophodontidae
Location:
North America (United States)
Length:
22 feet (6.5 meters)

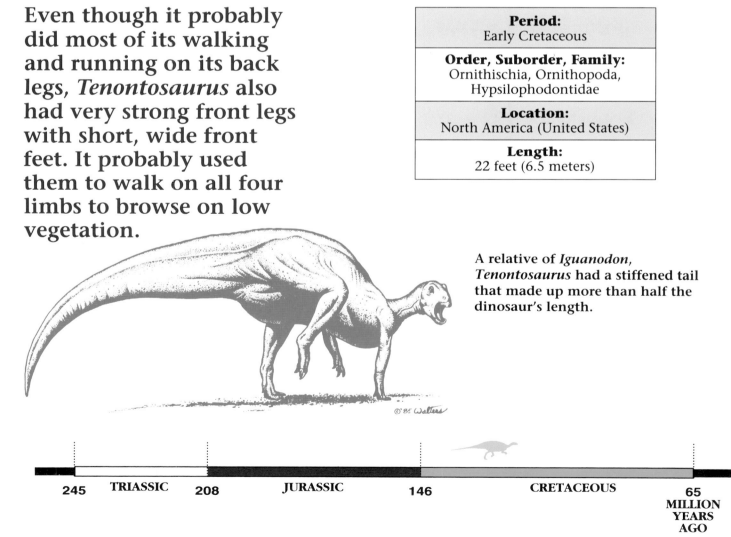

A relative of *Iguanodon*, *Tenontosaurus* had a stiffened tail that made up more than half the dinosaur's length.

| | 245 | TRIASSIC | 208 | JURASSIC | 146 | CRETACEOUS | 65 MILLION YEARS AGO |

ACROCANTHOSAURUS

(AK-roh-KANTH-oh-SORE-us)

Period:
Early Cretaceous

Order, Suborder, Family:
Saurischia, Theropoda, Allosauridae

Location:
North America

Length:
25 feet (7.5 meters)

A skeletal drawing of *Acrocanthosaurus*. The dinosaur had a large head, short arms, and a long slender tail that balanced the body when it ran.

Acrocanthosaurus had a unique feature—a tall "sail" along its neck, back, and tail. The sail was formed by very tall spines on the bones of the spine. Some of these spines were more than a foot tall; the spines of *Tyrannosaurus* were only half that tall.

Scientists have debated why *Acrocanthosaurus* had a sail along its back. Some think it released heat when the animal was too hot. It was probably used for display to make the animal look bigger when it faced a rival for territory or a mate.

No complete skeleton of *Acrocanthosaurus* has been found. Instead, paleontologists have built the dinosaur from parts of three skeletons. One skeleton had a three-foot-long skull.

Scientists have found footprints probably made by *Acrocanthosaurus* in several places in Texas. In one place, it looks as if *Acrocanthosaurus* stalked a large sauropod across a mud flat. When the sauropod footprints changed direction, so did those of *Acrocanthosaurus*. The outcome of the chase is not known; the end of the trackway has never been found.

There are other tall-spined meat-eating dinosaurs from Europe and Africa, but paleontologists do not know how *Acrocanthosaurus* is related to them. *Altispinax* had spines almost three feet tall, and *Spinosaurus* had spines six feet tall.

An adult *Acrocanthosaurus* was about ten feet tall at the hips and weighed up to three tons.

No complete skeleton of *Acrocanthosaurus* exists. Instead, paleontologists have built the dinosaur from parts of three skeletons.

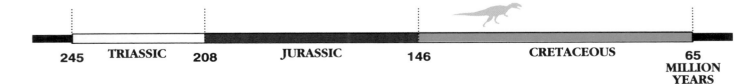

| 245 | TRIASSIC | 208 | JURASSIC | 146 | CRETACEOUS | 65 MILLION YEARS AGO |

DEINONYCHUS
(die-NON-ick-us)

At least three nearly complete skeletons of this small, fierce theropod were discovered in southern Montana in 1964. *Deinonychus* is the best-known member of the family Dromaeosauridae.

John Ostrom studied this dinosaur. Before his research, theropods were divided into two major groups: *Carnosauria* and *Coelurosauria*. Ostrom showed that *Deinonychus* had features of both groups. This helped convince other paleontologists that this division was incorrect. Also, several features of *Deinonychus* are also found in birds. Several paleontologists now think that *Deinonychus* and other dromaeosaurids are closely related to birds.

The thigh bone of *Deinonychus* was shorter than the shin bones, which shows the animal was a fast runner. As in other theropods and birds, the back foot was tridactyl, meaning that *Deinonychus* walked on only three toes (the three middle ones). But it had something completely different from other dinosaurs—a claw on the second toe that was very large, sharply pointed, and strongly curved. It is this "terrible claw" that the genus is named after. Studies of the sharp-clawed second toe show that it was not used while walking, but was carried off the ground in a raised position.

Another interesting feature of this genus and the family Dromaeosauridae is the thin rods of bone along the sides of the tail bones. These rods stiffened the back of the tail while allowing some flexibility.

Deinonychus may have grasped its prey with its jaws and hands while kicking the victim's underbelly with its large, knifelike claw.

Period:
Early Cretaceous
Order, Suborder, Family:
Saurischia, Theropoda, Dromaeosauridae
Location:
North America (United States)
Size:
8–10 feet (2.5–3 meters)

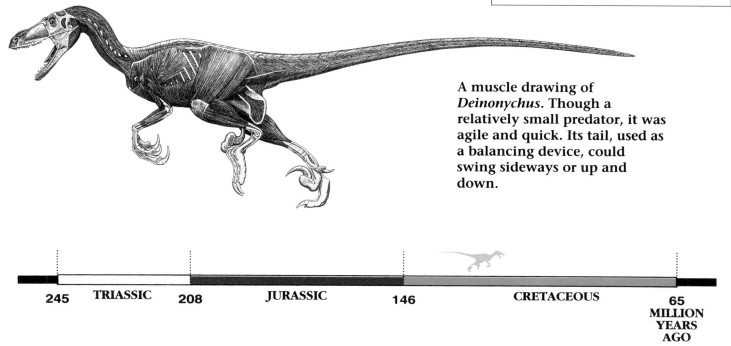

A muscle drawing of *Deinonychus*. Though a relatively small predator, it was agile and quick. Its tail, used as a balancing device, could swing sideways or up and down.

| 245 | TRIASSIC | 208 | JURASSIC | 146 | CRETACEOUS | 65 MILLION YEARS AGO |

MUTTABURRASAURUS

(MUTT-ah-BURR-ah-SORE-us)

Muttaburrasaurus langdoni. The dinosaur was only slightly smaller than *Iguanodon* from Europe and about the size of *Ouranosaurus* from western Africa. It is probably related to these two.

The "reptile from Muttaburra," *Muttaburrasaurus* is a recently discovered ornithopod from Australia—and one of the best known from that continent.

Muttaburrasaurus probably walked and ran on its back legs, but rested on all fours. A complete hand has not been found, so it is not known whether it had a thumb spike like *Ouranosaurus* and *Iguanodon*. Further discoveries will tell us more about this important part of the *Muttaburrasaurus* skeleton.

The broad head of *Muttaburrasaurus* looked like many other large ornithopods of its time. It did, however, have a large, hooked arch over the snout, unlike any other large ornithopods except the hook-nosed hadrosaurids. The front of the jaws was toothless and had a horny covering, much like turtles and birds. This beak was used to tear leaves and fruits from shrubs and low tree branches. The backs of the jaws were lined with many large teeth to tear apart food. The stomach of *Muttaburrasaurus* seems to have been large; it probably needed a large gut to digest a diet of plants.

Muttaburrasaurus and its relatives *Iguanodon*, *Ouranosaurus*, and *Probactrosaurus* show that these large ornithopods were worldwide. They are best known from the Early Cretaceous, before duckbilled dinosaurs appeared. When the duckbilled dinosaurs arose, there was a decline in all other ornithopods, among them *Muttaburrasaurus*.

Period:
Early Cretaceous
Order, Suborder, Family:
Ornithischia, Ornithopoda, Iguanodontidae
Location:
Australia
Length:
23 feet (7 meters)

Muttaburrasaurus probably walked and ran on its back legs, but rested on all fours.

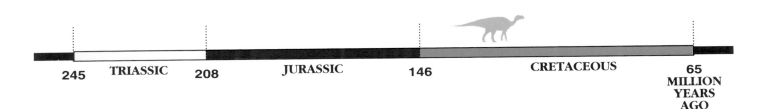

245	**TRIASSIC** 208	**JURASSIC** 146	**CRETACEOUS** 65 **MILLION YEARS AGO**

OURANOSAURUS

(oo-RAN-oh-SORE-us)

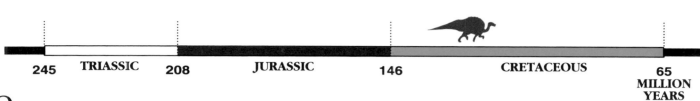

| 245 | TRIASSIC | 208 | JURASSIC | 146 | CRETACEOUS | 65 MILLION YEARS AGO |

Ouranosaurus ("brave reptile") resembled other large ornithopods. The dinosaur may have used all four legs to walk slowly or to rest. It probably used only its back legs for rapid running. Like its relatives *Iguanodon* and *Campto-saurus,* its hand had a sharp thumb spike that would have been a dangerous weapon against any predator.

The head of *Ouranosaurus* was large and its jaws were long. There was room for massive jaw muscles for chewing. The front of the snout was flat and broad and was probably covered by a horny beak. Above the eyes, along the top of the snout, was a pair of low, broad bumps. Only *Ourano-saurus* had these, and they may have been used for display, much like the small horns of some antelopes. A male *Ouranosaurus* might have used the bumps during head-butting contests, perhaps to defend territory, or the bumps may have been important for members of the family group or species to recognize each other.

Perhaps the most puzzling thing about *Ouranosaurus* is the large bony sail on its back. How this sail functioned has been somewhat perplexing. The sail may have helped *Ouranosaurus* regulate body temperature, with blood vessels' releasing or capturing heat depending on the dinosaur's needs. However, scientists are unsure.

> **. . . its hand had a sharp thumb spike that would have been a dangerous weapon against any predator.**

Period:
Early Cretaceous
Order, Suborder, Family:
Ornithischia, Ornithopoda, Iguanodontidae
Location:
Africa (Niger)
Length:
23 feet (7 meters)

Ouranosaurus nigeriensis. The dinosaur was a large and powerfully built animal that walked mostly on its stocky back legs. The front legs were smaller than the back, but were strongly built.

99

PSITTACOSAURUS

(sie-TACK-oh-SORE-us)

Psittacosaurus was one of the smallest and most primitive members of the Ceratopsia. It was found in rock that is thought to be Early Cretaceous (the exact age is not known). *Psittacosaurus* is the earliest known ceratopsian.

This small dinosaur did not look much like its later relatives. It had short arms and long grasping hands, and it walked on two legs. It also had a small head without a neck frill or horns. Although *Psittacosaurus* did not have a neck frill, it had a small, shelflike edge on the back of its skull that may have been the beginning of a frill. Without horns or a large frill for protection, *Psittacosaurus* probably escaped predators by running quickly.

Two of the specimens of *Psittacosaurus* were tiny juveniles; they are among the smallest known dinosaurs, even smaller than a robin. One tiny, nearly perfectly preserved skull is only about an inch long and would fit in a teaspoon. (A newly hatched *Psittacosaurus* would have been even smaller.) The teeth of the small *Psittacosaurus* specimens were slightly worn, which means they had already been feeding on tough plant material.

Psittacosaurus is the only known member of the family Psittacosauridae. Its closest relatives were the protoceratopsid dinosaurs *Protoceratops, Bagaceratops, Microceratops, Montanaceratops,* and *Leptoceratops.*

Period:
Early Cretaceous
Order, Suborder, Family:
Ornithischia, Ceratopsia, Psittacosauridae
Location:
Asia (Mongolian People's Republic, People's Republic of China, Siberia)
Length:
6½ feet (2 meters)

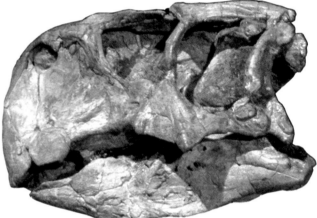

A *Psittacosaurus* skull. The beak inspired the name, which means "parrot reptile."

Psittacosaurus was first discovered in Outer Mongolia in 1922. Henry Osborn named the first *Psittacosaurus mongoliensis* specimen and gave a second specimen the name *Protiguanodon mongoliense.* Later, paleontologists realized that *Protiguanodon* and *Psittacosaurus* were the same animal.

Without horns or a large frill for protection, *Psittacosaurus* probably escaped predators by running quickly.

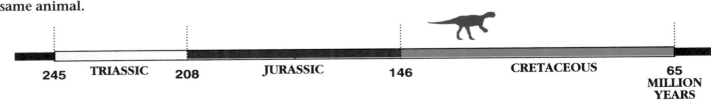

| 245 | TRIASSIC | 208 | JURASSIC | 146 | CRETACEOUS | 65 MILLION YEARS AGO |

SAUROPELTA
(SORE-oh-PEL-tuh)

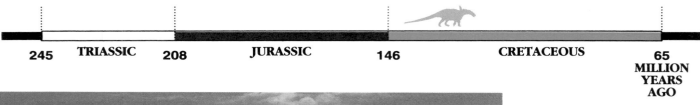

245	TRIASSIC	208	JURASSIC	146	CRETACEOUS	65 MILLION YEARS AGO

Sauropelta is known from several partial skeletons. These are important because much of the armor is in its natural position. From these, it has been possible to make the most accurate skeletal reconstructions and life restorations of any known ankylosaur.

Period:
Early Cretaceous

Order, Suborder, Family:
Ornithischia, Thyreophora, Nodosauridae

Location:
North America (United States)

Length:
17 feet (5 meters)

Like all members of the family Nodosauridae, *Sauropelta* had a long, tapering tail without the bony club of members of the family Ankylosauridae. The neck of *Sauropelta* had long bone spikes projecting up and out; such spikes are not found in ankylosaurids. These spikes would have protected the neck from the bite of predators, including *Acrocanthosaurus*. Perhaps more important, the spikes would have made the animal look bigger and more dangerous. Bluffing would have helped it avoid a fight.

The armor of *Sauropelta,* and all ankylosaurs, formed in the skin, just as it does in modern alligators, crocodiles, and certain lizards. It got its name from this feature; the name means "lizard skin." The armor of *Sauropelta* consisted of rows of oval plates across the neck, back, and tail; spines and spikes along the sides; and large circular plates over the hips. Tiny, irregular pieces of armor filled the gaps between the larger plates. Even the skull had armor, with the plates tightly joined to the outer surface of the skull and jaws.

Sauropelta was different from many ankylosaurs because it had two types of teeth. Small pegs lined the upper front of the mouth, while leaf-shaped teeth lined both the upper and lower cheek region. The shape of the cheek teeth and their pattern of wear show that *Sauropelta* ate soft plants.

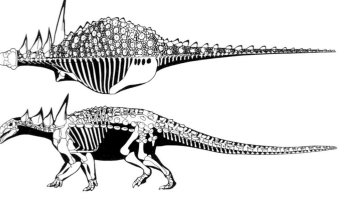

Skeletal drawings of *Sauropelta.* Note the bone spikes along the neck. Well armored, the dinosaur even had plates joined to skull and jaws.

101

WUERHOSAURUS

(woo-AIR-oh-SORE-us)

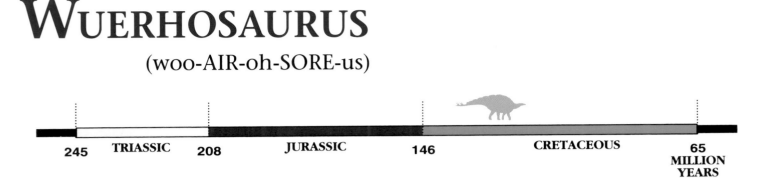

| 245 | TRIASSIC | 208 | JURASSIC | 146 | CRETACEOUS | 65 MILLION YEARS AGO |

As far as paleontologists know, stegosaurs almost became extinct at the close of the Jurassic Period. Very few lived during the Cretaceous Period. *Craterosaurus,* known only from a piece of a back bone, is from the Early Cretaceous of Great Britain. Another stegosaurian, *Dravidosaurus,* is known from remains from the Late Cretaceous of India. The undescribed stegosaurian "Monkonosaurus" from Tibet may also be of Early Cretaceous age. But the best-documented Early Cretaceous stegosaur is *Wuerhosaurus.*

At present, the latest known Chinese stegosaurian, *Wuerhosaurus homheni* is based on a fragmentary skeleton lacking the skull and on three tail bones of a second animal, all found in the Tugulo Formations near the northwestern part of the Junggar Basin. The dinosaur was described by Dong Zhiming in 1973. Two armor plates found with the skeleton are thin, long, low, and somewhat semicircular. They are quite different from the tall, triangular plates of other stegosaurians.

The dinosaur's body was broad, as shown by its wide pelvic bones, and its front limbs were quite short. Although the back limbs have not been found, the shortness of the front limbs shows that the animal had an arched back—perhaps even more curved than in *Stegosaurus.*

Period:
Early Cretaceous
Order, Suborder, Family:
Ornithischia, Thyreophora, Stegosauridae
Location:
Asia (People's Republic of China)
Length:
20 feet (6 meters)

The dinosaur's body was broad, as shown by its wide pelvic bones, and its front limbs were quite short. Although the back limbs have not been found, the shortness of the front limbs shows that the animal had an arched back. . . .

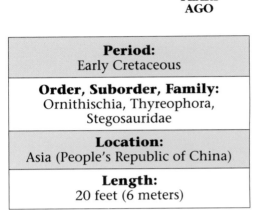

Wuerhosaurus homheni. The dinosaur seems closely related to *Stegosaurus* and, according to some paleontologists, may also have had alternating armor plates along its back.

CHAPTER 6
DINOSAURS RULE THE WORLD

Late Cretaceous Period

The continents continued to separate during the Late Cretaceous. North and South America moved west into the Pacific Ocean, and the North and South Atlantic Oceans widened. India continued to move north toward Asia. Still joined together, Australia and Antarctica journeyed away from Africa, going beyond the South Pole.

During the early Late Cretaceous, the climate was warm. As the period came to a close, the average climate became cooler, but it was still much warmer than today. By the end of the period, the tropics were only in areas near the equator. The climate in the farther northern and southern hemispheres (and polar regions) became temperate and more seasonal, with cool winters and warm summers. Forests in the temperate zones became less tropical, with magnolias, sassafras, redwoods, and willow trees plentiful.

Late Cretaceous map.

Dinosaurs remained the main large land animals. Smaller land animals included turtles, crocodiles, snakes, lizards, frogs, and salamanders. Mammals remained small in size, but mammals that gave birth to live young appeared for the first time.

Plant-eaters were eaten by fierce tyrannosaurids. Most of those known from North America were 25 to 35 feet long, such as *Albertosaurus*. But the Late Cretaceous saw one of the smallest, *Nanotyrannus* (about 18 feet long), and the largest, *Tyrannosaurus* (40 feet long). Ankylosaurs and nodosaurs were heavily armored plant-eaters that did not need the protection of a herd to avoid being eaten.

Albertosaurus confronting two styracosaurs. The Late Cretaceous was a time when plant-eating dinosaurs had much to fear from fierce tyrannosaurids.

Eastern Asia—especially Mongolia—was an abundant source of dinosaur species. Tyrannosaurids from Asia were smaller and more primitive than their North American relatives. Dome-headed dinosaurs from Mongolia were much different from those in North America. The most interesting dinosaurs discovered in Mongolia and China are the segnosaurs. Wide-bodied plant-eaters with powerful claws, such as *Erlikosaurus, Segnosaurus,* and *Therizinosaurus,* are known from nowhere else in the world.

Small predators closely related to those in western North America abounded in eastern Asia. These included the sickle-clawed dromaeosaurids *Adasaurus, Hulsanpes,* and *Velociraptor. Shanshanosaurus* from China may have been related to North America's *Aublysodon.*

Europe was covered by a continental sea that divided it into islands. This led to the evolution of dwarf dinosaurs known as "island endemics"—*Struthiosaurus, Magyarosaurus, Hypselosaurus, Rhabdodon,* and *Craspedodon.*

The Gondwanaland continents had different Late Cretaceous dinosaurs. Almost all the Late Cretaceous Gondwanaland sauropods were titanosaurids. They were found mainly in South America, Africa, India, and Madagascar.

The Late Cretaceous Period ended about 65 million years ago. All dinosaurs, pterosaurs, and certain marine reptiles vanished from the face of the earth. It seems that this extinction happened at almost the same time on all continents. The rule of the dinosaurs came to a close with the end of the Mesozoic Era.

A dromaeosaur feeding on the carcass of *Chasmosaurus*. The Late Cretaceous was a time when small predators related to those in western North America abounded in eastern Asia.

A Late Cretaceous scene, with the duckbilled dinosaur *Corythosaurus* in the foreground.

ABELISAURUS
(AH-bell-ih-SORE-us)

| 245 | TRIASSIC | 208 | JURASSIC | 146 | CRETACEOUS | 65 MILLION YEARS AGO |

During the Cretaceous, dinosaurs that lived in the southern hemisphere were much different from their northern-hemisphere relatives. This was the result of the separation of the northern and southern land masses that began in the Jurassic Period. The recently discovered large theropod *Abelisaurus comahuensis,* from Patagonia, Argentina, looked a little like *Albertosaurus* from Alberta, Canada, particularly in its size and lifestyle. But parts of its skull led two Argentinian paleontologists to put it in its own family, the Abelisauridae. They think it was more closely related to the theropod *Ceratosaurus* from the Jurassic Period.

| **Period:** |
| Late Cretaceous |
| **Order, Suborder, Family:** |
| Saurischia, Theropoda, Abelisauridae |
| **Location:** |
| South America (Argentina) |
| **Length:** |
| 25–30 feet (7.5–9 meters) |

Only the skull of *Abelisaurus* has been found, but its body proportions were probably similar to other large theropods with the same size skulls (three feet long). *Carnotaurus* had slender legs with the front shorter than the back. Since *Carnotaurus* probably was an early abelisaurid, scientists suppose that *Abelisaurus* also had short front limbs and slender legs. Other details of its body are unknown.

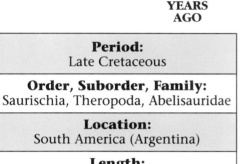

The discovery of *Abelisaurus* is important because it has shed light on many different southern-hemisphere theropods that are known only from fragmentary and puzzling material. These remains were difficult to identify and were occasionally used to suggest that Late Cretaceous tyrannosaurids from the northern hemisphere were in the southern hemisphere. Now that good abelisaurid material has been discovered and described, scientists have learned that many of those remains were abelisaurids. The possibility of southern-hemisphere tyrannosaurids is less likely.

The discovery of *Abelisaurus,* which was related to the theropod *Ceratosaurus* from an earlier period, is important. The dinosaur sheds light on various southern-hemisphere theropods about which little is known.

. . . scientists suppose that *Abelisaurus* . . . had short front limbs and slender legs.

ALBERTOSAURUS
(al-BUR-toh-SORE-us)

*A*lbertosaurus was an older "cousin" to the better-known *Tyrannosaurus*. In many ways the two were similar: The head was large compared to the body, the tiny forearms had only two fingers each, and the long tail balanced the body over two powerful back legs. But the eyes of *Tyrannosaurus* looked forward, while those of *Albertosaurus* looked more toward the side. This suggests that *Albertosaurus* did not judge distances as well, so that when it hunted it probably did not leap onto its prey.

A skeletal drawing of *Albertosaurus*, a relative of *Tyrannosaurus*. Fossil remains of the dinosaur, especially its teeth, are numerous.

Stealth, power, and speed were its biggest assets. With its long, powerful rear legs, *Albertosaurus* could outrun its prey or ambush a heavy plant-eater that stood alone and unprotected. The rear legs could deliver crushing blows, knocking the prey off balance. *Albertosaurus* delivered deadly wounds with its claws. The light build and long legs show that it was fast and graceful. It may have been able to run 25 to 30 miles per hour.

The head of *Albertosaurus* had two small, blunt horns, just in front of the eyes. These may have been for show, much like the comb on a chicken today. It is possible that the male had brightly colored skin covering the horns to attract the female during mating season, much as birds today, with the males brightly colored to attract females.

Fossil remains of *Albertosaurus* are numerous, especially teeth, which often broke when it was feeding. Several species are recognized: *Albertosaurus sarcophagus* and *Albertosaurus libratus* are the most common. *Albertosaurus lancensis* has been recently renamed *Nanotyrannus*. Some paleontologists think the theropod dinosaur *Alectrosaurus olseni* from Mongolia is a species of *Albertosaurus*. If this is correct, *Albertosaurus* lived in both North America and Asia.

Period:
Late Cretaceous
Order, Suborder, Family:
Saurischia, Theropoda, Tyrannosauridae
Location:
North America
Length:
30 feet (9 meters)

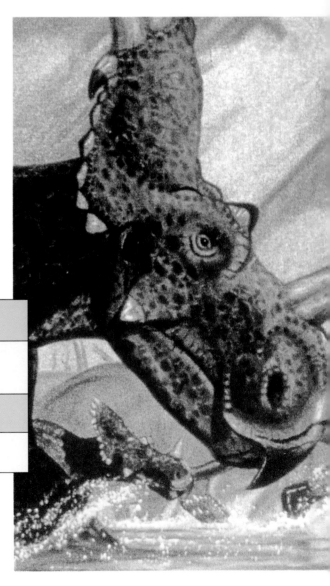

Albertosaurus bothering a herd of styracosaurs. It was a dangerous predator despite its relatively light build.

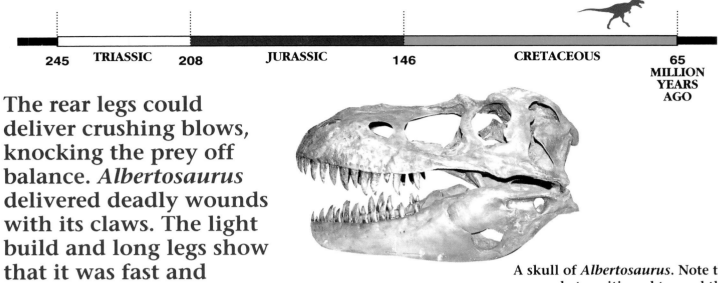

| 245 | TRIASSIC | 208 | JURASSIC | 146 | CRETACEOUS | 65 MILLION YEARS AGO |

The rear legs could deliver crushing blows, knocking the prey off balance. *Albertosaurus* delivered deadly wounds with its claws. The light build and long legs show that it was fast and graceful. It may have been able to run 25 to 30 miles per hour.

A skull of *Albertosaurus*. Note the eye socket positioned toward the side and the powerful jaws with cutting teeth. In life, the dinosaur had two small, blunt horns in front of the eyes, possibly colored brightly in males.

ANKYLOSAURUS

(an-KIE-loh-SORE-us)

*A*nkylosaurus is one of the best-known dinosaurs. So, it may come as a surprise that this dinosaur is known only from three partial skeletons, none of which has been fully described. Its popularity mostly comes from a fanciful life-size restoration made for the 1964 New York World's Fair.

Through the studies of Walter Coombs, we now know that *Ankylosaurus* and all members of the family Ankylosauridae did not have spines and spikes projecting from the body as shown in the World's Fair restoration. Such spines and spikes only occur in the other ankylosaur family, Nodosauridae (see *Sauropelta*).

There is no basis for showing keeled, or ridged, rectangular plates in even rows across the body. Keeled plates have been found with the *Ankylosaurus* specimens, but these are different sizes and shapes. Scientists do not know the arrangement of the plates on the body because no specimen has been found with the plates preserved as they were in life. The arrangement of plates is known for the nodosaurids *Sauropelta* and *Edmontonia* and the ankylosaurid *Saichania*. These specimens are important because they provide the only proof of how the armor looked in various species of ankylosaurs.

Like most armored dinosaurs, *Ankylosaurus* had bone plates joined to the outside of the skull and jaws. But unlike *Sauropelta*, which had many large plates, *Ankylosaurus* had many small ones. *Ankylosaurus* also differs from *Sauropelta* in having hornlike projections in the upper and lower corners of the skull behind the eyes. Why *Ankylosaurus,* and all ankylosaurids, developed these horns is not understood. They may have been used in combat, one ankylosaur against another. The animals would have stood side by side and swung their heads into each other's body. Such a blow would be painful but not fatal.

Ankylosaurus had a large bone club on the end of its tail. All members of its family had these clubs, and the lack of a club characterizes the other ankylosaur family, the Nodosauridae. The shape of the club is different for each family member. The club of *Ankylosaurus* was wide and long. The club in all the ankylosaurids was made of large armor plates fused together at the end of the tail. The bones of the tail were modified for swinging the club: They interlocked to form a "handle," allowing *Ankylosaurus* to put force into the swing.

An *Ankylosaurus* skull. The dinosaur, known only from three partial skeletons, had hornlike projections in the upper and lower corners of the skull behind the eyes and bone plates joined to the skull and jaws.

Ankylosaurus had ridged armor plates and, typical of ankylosaurids, a clubbed tail. It did not have spikes or spines, as was once thought.

108

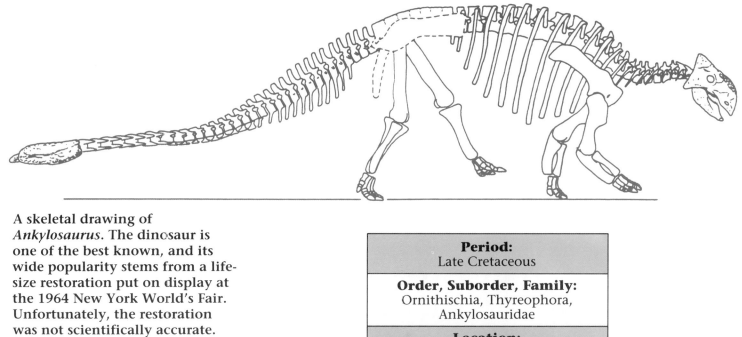

A skeletal drawing of *Ankylosaurus*. The dinosaur is one of the best known, and its wide popularity stems from a life-size restoration put on display at the 1964 New York World's Fair. Unfortunately, the restoration was not scientifically accurate.

Period:	
Late Cretaceous	
Order, Suborder, Family:	
Ornithischia, Thyreophora, Ankylosauridae	
Location:	
North America	
Length:	
23 feet (7 meters)	

The bones of the tail were modified for swinging the club: They interlocked to form a "handle," allowing *Ankylosaurus* to put force into the swing.

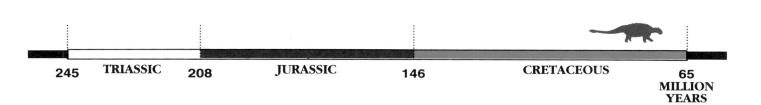

| 245 | TRIASSIC | 208 | JURASSIC | 146 | CRETACEOUS | 65 MILLION YEARS AGO |

CHASMOSAURUS

(KAZ-moh-SORE-us)

The first *Chasmosaurus* fossil found was part of the neck frill. It was unearthed in 1898 by Lawrence Lambe along the Red Deer River, Alberta, Canada. Lambe first thought it belonged to the genus *Monoclonius*, so he named it *Monoclonius belli*. In 1914, after a more complete skull had been found by Charles Sternberg, Lambe realized that his original find was a new ceratopsian and renamed it *Chasmosaurus belli*. Since then, more skulls and skeletons of *Chasmosaurus* have been found.

Chasmosaurus had a very long frill with enormous openings on its surface. The outline of the frill is indented, making it look heart-shaped when seen from above. It is the enormous, or cavernous, frill holes for which this dinosaur is named; its name means "chasm reptile."

Chasmosaurus had a short nasal horn and brow horns that were of different lengths; some species had long brow horns while others had short ones. One specimen collected by Sternberg had impressions of *Chasmosaurus* skin. These show that the rough skin of *Chasmosaurus* had large, circular bumps in regularly spaced rows, with many smaller bumps between them.

Chasmosaurus was most closely related to *Anchiceratops*, *Arrhinoceratops*, *Torosaurus*, *Triceratops*, and *Pentaceratops*. *Chasmosaurus* was probably the oldest and the smallest of all chasmosaurine ceratopsians.

Chasmosaurus, first discovered in 1898 in Alberta, Canada. Its facial horns varied in length, depending on the species and sex. The dinosaur is also known for its long, indented frill and regularly spaced row of bumps on its skin.

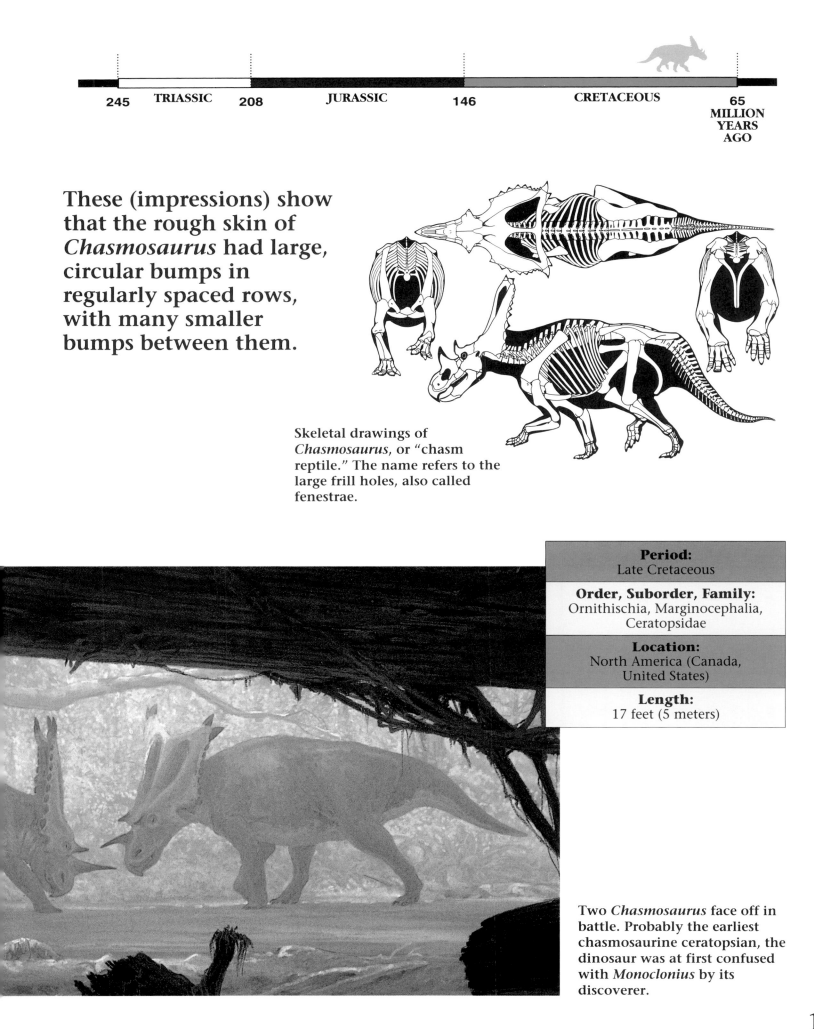

245	TRIASSIC	208	JURASSIC	146	CRETACEOUS	65 MILLION YEARS AGO

These (impressions) show that the rough skin of *Chasmosaurus* had large, circular bumps in regularly spaced rows, with many smaller bumps between them.

Skeletal drawings of *Chasmosaurus*, or "chasm reptile." The name refers to the large frill holes, also called fenestrae.

Period:	Late Cretaceous
Order, Suborder, Family:	Ornithischia, Marginocephalia, Ceratopsidae
Location:	North America (Canada, United States)
Length:	17 feet (5 meters)

Two *Chasmosaurus* face off in battle. Probably the earliest chasmosaurine ceratopsian, the dinosaur was at first confused with *Monoclonius* by its discoverer.

111

DRYPTOSAURUS
(DRIP-toh-SORE-us)

Dryptosaurus is the only meat-eating dinosaur from the East Coast of the United States that is based on more than a single bone. The partial skeleton was discovered more than a hundred years ago by workers in a quarry in New Jersey. It was originally named *Laelaps,* but this name had been given to a spider, so the dinosaur was renamed.

The only parts of the skull that have been found of *Dryptosaurus* are pieces of the jaws. The teeth had jagged edges like those on a steak knife, showing it was a meat-eater. This is supported by the huge, eight-inch claw (it probably had several, but only one was found). Hands with talons like an eagle's would have helped *Dryptosaurus* hold struggling prey. Its name means "tearing lizard," which refers to these claws.

Although in fragments, enough of the skeleton has been found to show that this animal stood about eight feet tall at the hips. The back legs were much longer than the front, so *Dryptosaurus* probably walked on its back legs with its tail acting as a balance.

The "tearing lizard" was aptly named after a huge fossil claw. Originally, *Dryptosaurus* was called *Laelaps*, but this was changed after scientists realized that name already belonged to a spider.

Dryptosaurus is a puzzle because its relationship to other carnivorous dinosaurs is not known. Although it looked like a tyrannosaur, it had longer limbs for its size and larger, more curved claws on its hands. Scientists hope it will be better understood when they have finished restudying the skeleton.

It is not certain what *Dryptosaurus* ate, because the fossil record for dinosaurs in New Jersey is very incomplete. All dinosaur fossils from New Jersey, including both *Dryptosaurus* and the older *Hadrosaurus,* were found in marine rocks. The remains of these dinosaurs must have drifted out to sea before becoming fossilized.

Hadrosaur bones are in many Late Cretaceous formations in the area, so hadrosaurs probably made up a large part of the diet of *Dryptosaurus.* Ceratopsians were probably not present (none of their distinctive bones have been found), so *Dryptosaurus* did not have to dodge the horns of an angry ceratopsian. Only isolated bone armor plates have been found from nodosaurid ankylosaurs. The body armor of ankylosaurs would have made it difficult for *Dryptosaurus* to kill the animal, but it is possible that *Dryptosaurus* scavenged an already dead body. *Dryptosaurus* may have been like the African lion that feeds on a carcass rather than chasing an antelope.

. . . it is possible that **Dryptosaurus** scavenged an already dead body. **Dryptosaurus** may have been like the African lion that feeds on a carcass rather than chasing an antelope.

DRYPTOSAURUS

245	TRIASSIC	208	JURASSIC	146	CRETACEOUS	65 MILLION YEARS AGO

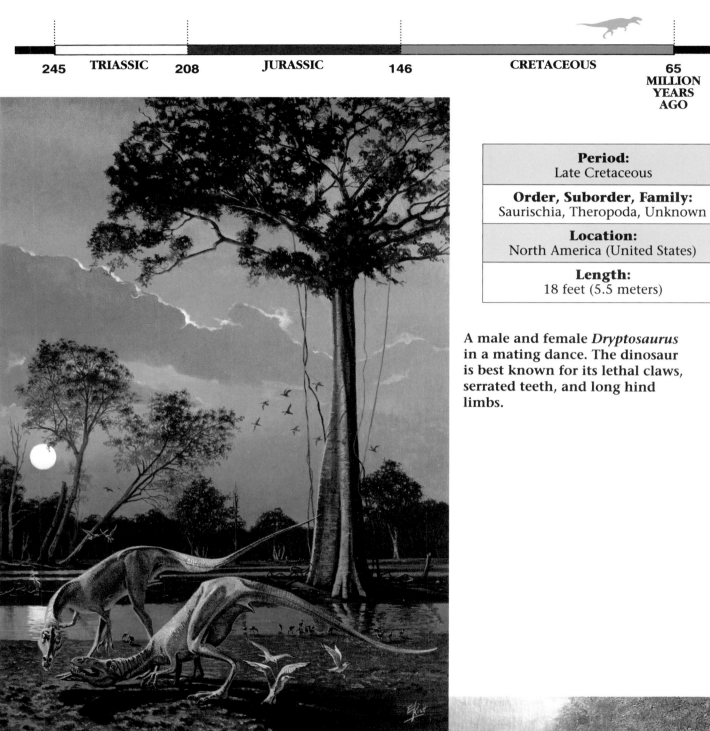

Period:
Late Cretaceous
Order, Suborder, Family:
Saurischia, Theropoda, Unknown
Location:
North America (United States)
Length:
18 feet (5.5 meters)

A male and female *Dryptosaurus* in a mating dance. The dinosaur is best known for its lethal claws, serrated teeth, and long hind limbs.

Two dryptosaurs attacking a hadrosaur. Though it's not certain what *Dryptosaurus* ate, hadrosaur bones have been found in many formations associated with the Late Cretaceous predator.

EDMONTOSAURUS

(ed-MON-toh-SORE-us)

Edmontosaurus was one of the largest hadrosaurids. This flat-headed duckbilled dinosaur was originally found, described, and named by Lawrence Lambe in 1920. It was one of the last dinosaurs, surviving to the end of the Mesozoic Era along with *Tyrannosaurus* and *Triceratops.*

Like all hadrosaurids, *Edmontosaurus* had a broad snout, long jaws, and large eyes. The snout probably had a horny covering that closed against the horny covering of the lower jaw. This was how the animal bit off leaves and branches from shrubs and low-lying boughs. To chew the food, the animal had many teeth behind its beak that were arranged in a grinding pattern. The teeth plus the many replacement teeth (there were often a thousand or more in the jaws) formed what is called the dental battery. *Edmontosaurus* and all other hadrosaurids were able to move their jaws slightly from side to side. This made them successful plant-eaters.

The nostrils of *Edmontosaurus* were large and hollow. They may have been covered with loose skin that the animal could have filled with air. When inflated, these bags may have been used to make loud bellowing sounds. Communication was probably important to this animal; it lived in large groups with both adults and offspring. These bags may also have been brightly colored for display to attract a mate or to help animals from the same species recognize each other.

Edmontosaurus is responsible for the most spectacular of all dinosaur fossils. These are the famous hadrosaur "mummies"—skeletons found with skin impressions around the head, shoulders, arms, legs, and tail. These impressions show the pattern of bumps, called tubercles, that were found on the skin.

The name *Edmontosaurus* means the "reptile from Edmonton," signifying the provincial capital Alberta where it was found. It was related to other flat-headed hadrosaurids, including *Shantungosaurus* from the People's Republic of China. It was not as closely related to the solid-crested hadrosaurids, such as *Saurolophus* and *Prosaurolophus,* which lived in the same areas as *Edmontosaurus.*

After much study, *Anatosaurus* has been found to be the same animal as *Edmontosaurus,* which was its first and therefore correct name.

It was one of the last dinosaurs, surviving to the end of the Mesozoic Era along with *Tyrannosaurus* and *Triceratops.*

An *Edmontosaurus* skeletal drawing. Related to China's *Shantungosaurus* and other flat-headed hadrosaurids, the "reptile from Edmonton" had a thousand or more replacement teeth arranged in a grinding pattern.

EDMONTOSAURUS

245	**208**	**146**	**65 MILLION YEARS AGO**
TRIASSIC	JURASSIC	CRETACEOUS	

A pair of *Edmontosaurus annectens* skeletons.

An *Edmontosaurus* skin impression. The famous hadrosaur "mummies" are skeletons with skin impressions showing the pattern of skin bumps, or tubercles.

Edmontosaurus may have been able to make bellowing sounds by inflating skin-covered nostrils. It probably bit off vegetation with the help of a horny covering on its snout and lower jaw.

Period:
Late Cretaceous
Order, Suborder, Family:
Ornithischia, Ornithopoda, Hadrosauridae
Location:
North America (Canada, United States)
Length:
42 feet (13 meters)

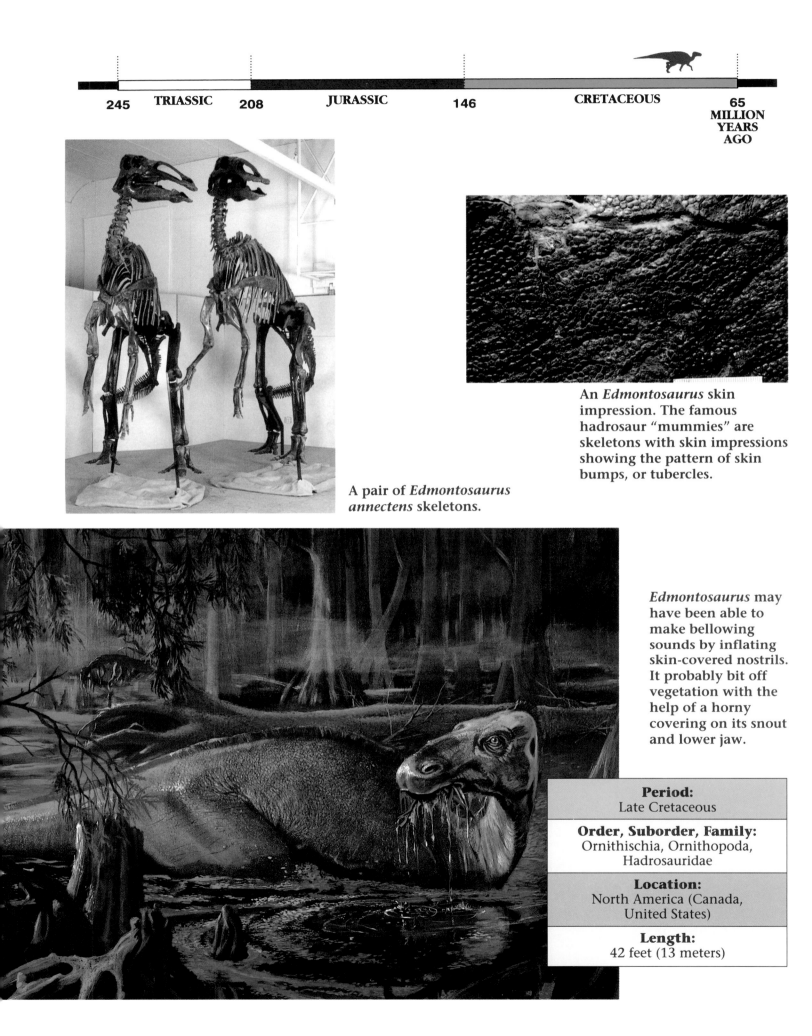

MAIASAURA

(MY-ah-SORE-ah)

Since it was named in 1979 by John Horner and Robert Makela, *Maiasaura* has become one of the most famous dinosaurs, especially in the ways it cared for its young and for what it tells us of the early development of dinosaurs.

The name *Maiasaura* means "good mother reptile." This large dinosaur laid her eggs in a mound of dirt in a circular or spiral pattern. Then she covered the mound with vegetation, much as crocodiles do today. The eggs were warmed by the rotting vegetation. Rather than sit directly on the nest, mother *Maiasaura* probably sat next to it to keep thieves from stealing the eggs and other maiasaur parents from walking on them. Because females of a herd made nests in the same area, this would have been a danger.

When the eggs hatched, the hatchlings were too immature to leave the nest. Their limbs were not well developed, but there was wear on their teeth. This means the hatchlings stayed in their nests and their parents brought them food. According to Horner, the young *Maiasaura* probably lived in nests until they were a year or two old. They grew from about 16 inches to 58 inches in only a year; this is rapid growth and may mean that they were warm-blooded.

A hatchling *Maiasaura* had a tall, narrow head. As it grew, its head became lower and wider from back to front, with a broad, horny beak. Over its eyes was a stout, bony crest that probably was used for display and head-butting during breeding season or when defending territory. Because these dinosaurs lived in very large groups, with up to 10,000 animals in a herd, they needed to protect their territory.

All the *Maiasaura* specimens come from the Late Cretaceous Two Medicine Formation of western Montana. *Orodromeus* also lived in this area at the same time, as well as the predators *Albertosaurus* and *Troodon*. *Maiasaura* has not been found in other areas of the West.

We know more about the life of *Maiasaura* than any other dinosaur; we also know a lot about its evolutionary relationships. The closest relative

of *Maiasaura* was *Brachylophosaurus* from southern Alberta and Montana. More remotely, these two solid-crested hadrosaurids were related to the more primitive *Hadrosaurus*, *Aralosaurus*, and *Gryposaurus*.

Baby *Maiasaura* skeletons. Hatched from nests that were warmed by rotting vegetation, young maiasaurs were probably fed by their parents at first.

MAIASAURA

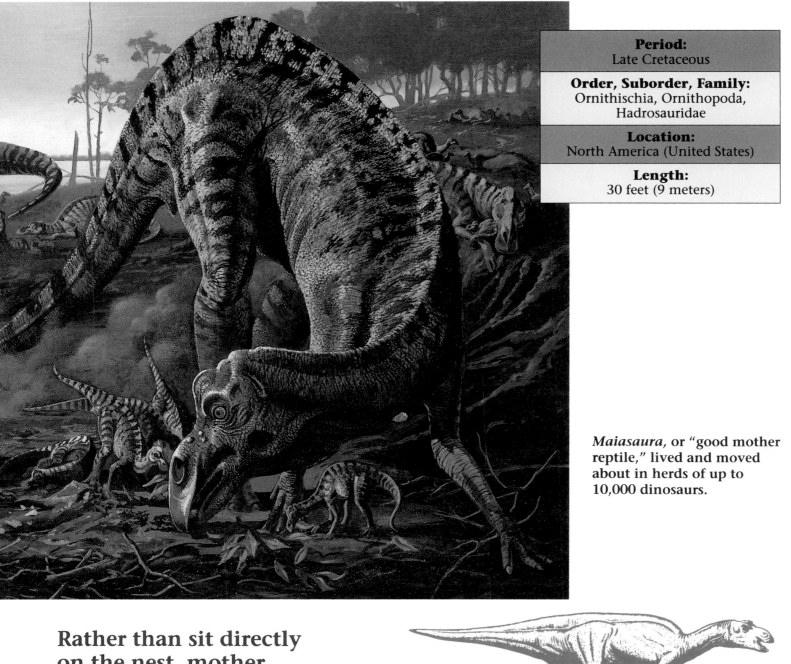

Period:
Late Cretaceous
Order, Suborder, Family:
Ornithischia, Ornithopoda, Hadrosauridae
Location:
North America (United States)
Length:
30 feet (9 meters)

Maiasaura, or "good mother reptile," lived and moved about in herds of up to 10,000 dinosaurs.

Rather than sit directly on the nest, mother *Maiasaura* probably sat next to it to keep thieves from stealing the eggs and other maiasaur parents from walking on them.

Maiasaura is known only from a Late Cretaceous formation in western Montana, but scientists have learned a great deal about the dinosaur and its evolutionary relationships.

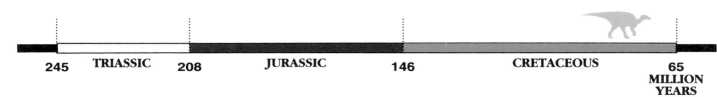

| 245 | TRIASSIC | 208 | JURASSIC | 146 | CRETACEOUS | 65 MILLION YEARS AGO |

PROTOCERATOPS
(PROH-toh-SAIR-ah-tops)

One of the most celebrated dinosaurs of the 20th century, *Protoceratops andrewsi* was discovered in Mongolia in 1922 by an expedition from the American Museum of Natural History led by Roy Chapman Andrews. Its genus means "first horned face," and its species name was in honor of the expedition's leader. Workers also discovered the skeleton of a nimble, toothless predator that was later named *Oviraptor philoceratops*, or "egg-stealer, lover of ceratopsians." In the Gobi Desert were parents, nests, eggs, hatchlings, and predators, all in one amazing deposit. This was the first discovery of dinosaur eggs, and the discovery made news everywhere.

Period:	Late Cretaceous
Order, Suborder, Family:	Ornithischia, Marginocephalia, Protoceratopsidae
Location:	Asia (Mongolian People's Republic)
Length:	6 feet (1.8 meters)

Protoceratops was a small, compact dinosaur, only six feet long and two feet tall at the hips as an adult. It weighed less than 400 pounds. Hatchlings were only a foot long, and the eggs were eight inches long and seven inches around. Although small as an adult, *Protoceratops* had a sturdy build. The front legs were nearly as long as the back so that it could carry its heavy head and jaws. Its toes had claws that it used to dig in low vegetation for leaves and twigs. Its heavy tail balanced the animal when it walked.

The dinosaur had a powerful bite; it cropped low vegetation with its beak. Behind its beak, it had dozens of teeth that chopped the tough leaves and branches into smaller pieces, perhaps in a chewing motion. Food collected in its fleshy cheeks on the sides of the jaws. Along with low-growing shrubs and trees, *Protoceratops* may also have eaten the newly evolving flowering plants that appeared in the Late Cretaceous.

Protoceratops did not have horns on its face or a shield like those found on its relative *Triceratops*. *Protoceratops* did have a slight bump on the snout below the eyes that may have been the beginning of a horn. The bump was larger on males, which also had larger frills. Males may have used these features to attract females. With the discovery of so many skeletons, scientists concluded that *Protoceratops* lived in herds.

Relatives of *Protoceratops* included *Montanaceratops* and *Leptoceratops* from North America and *Bagaceratops* and *Microceratops* from Asia. All are from the Late Cretaceous and are similar to their more famous cousin. There may have been land connections between the northern continents in the Late Cretaceous that allowed these small ceratopsians to migrate.

A *Protoceratops* skull. The dinosaur had a parrotlike beak that was rounded and toothless. The rear of the skull was expanded into great ridges of bone where the jaw muscles attached. The bony frill was small and simple in *Protoceratops*, but became large and elaborate in some of its giant relatives. The sides of the frill were open, which lightened the skull without sacrificing strength.

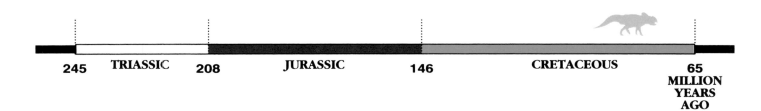

| 245 | TRIASSIC | 208 | JURASSIC | 146 | CRETACEOUS | 65 MILLION YEARS AGO |

A surprise for paleontologists was the discovery that *Protoceratops* lived in the desert. The eggs were laid in shallow holes in the sand. Also, the small pores and the pattern on the shells show that the eggs were adapted for desert conditions.

Protoceratops may be the ancestor of all horned dinosaurs. Because of this evolutionary link, and because of all the fossils found, it has been the subject of much study. Like all ceratopsians, *Protoceratops* was from the Late Cretaceous, but it was older than most of its relatives.

Protoceratops was a small, compact dinosaur, only six feet long and two feet tall at the hips as an adult. It weighed less than 400 pounds. Hatchlings were only a foot long, and the eggs were eight inches long and seven inches around.

Eggs thought to be from a *Protoceratops* nest in the Gobi Desert. From the same area have come an amazing number of fossils, including the skeleton of *Oviraptor*, *Protoceratops*'s predator.

TRICERATOPS

(trie-SAIR-ah-TOPS)

Triceratops is one of the most spectacular and well known of all dinosaurs. This huge ceratopsian, with its long, pointed brow horns and curving neck frill, was among the last dinosaurs to walk the earth and lived to the end of the Cretaceous.

The body of *Triceratops* was massive with a huge, barrellike rib cage and short tail. Except for its large size, it probably looked like the other large ceratopsids. The dinosaur was probably not a fast animal. It had heavily built limb bones, and the front limbs were shorter than the back. It probably relied on strength rather than speed for defense.

Although the name *Triceratops* means "three-horned face," not all specimens had three horns. Often the nasal horn was either very short or nearly absent. However, the two brow horns, which grew out of the top of the skull over each eye, were always large and well developed.

The frill of *Triceratops* was different from all other ceratopsids. It was broad and round, and the bone was thick and had no openings. Some paleontologists believe that the broad frill, with a large blood supply, may have helped heat or cool the animal.

Like the horns, the solid frill may have been used to protect *Triceratops* from predators or from other *Triceratops*. The frill may also have been for bluff; *Triceratops* could put its head down and point its brow horns at another animal, making the frill stand up. The frill may also have helped attract a mate.

For many years, paleontologists thought that *Triceratops* was most closely related to the centrosaurine (or "short-frilled") ceratopsids because the frill of *Triceratops* was relatively short. However, since other features of the skull are like those of the Chasmosaurinae, such as long brow horns, a short nasal horn, and a long snout with a double opening of the nose, *Triceratops* is now considered a chasmosaurine.

Triceratops had a very long and powerful beak. Each jaw had closely packed teeth with a broad grinding surface.

Triceratops had a small brain; the ratio of its brain size to its body size is lower than two-legged dinosaurs, such as the duckbills and meat-eaters.

Period:
Late Cretaceous
Order, Suborder, Family:
Ornithischia, Marginocephalia, Ceratopsidae
Location:
North America (Canada, United States)
Length:
30 feet (9 meters)

A *Triceratops* skull. With its scissorlike beak and grinding teeth, the dinosaur was able to bite off and chew even the toughest plants. Because its beak was narrow and pointed, it probably bit off plants with the side of its beak. Muscular cheeks held food in its mouth as it chewed.

The dinosaur was probably not a fast animal. It had heavily built limb bones, and the front limbs were shorter than the back. It probably relied on strength rather than speed for defense.

The world in which *Triceratops* lived looked quite contemporary. The landscape had modern-looking trees and shrubs. Animals that lived at the same time as *Triceratops* included *Tyrannosaurus*, *Thescelosaurus*, *Torosaurus*, *Leptoceratops*, *Ankylosaurus*, and *Ornithomimus*.

Some *Triceratops* specimens have healed wounds in their skull or frill, showing that they battled among themselves. They may have fought over females, territory, or leadership. It is also likely that the long, strong horns of *Triceratops* were used as weapons against predators such as *Tyrannosaurus*.

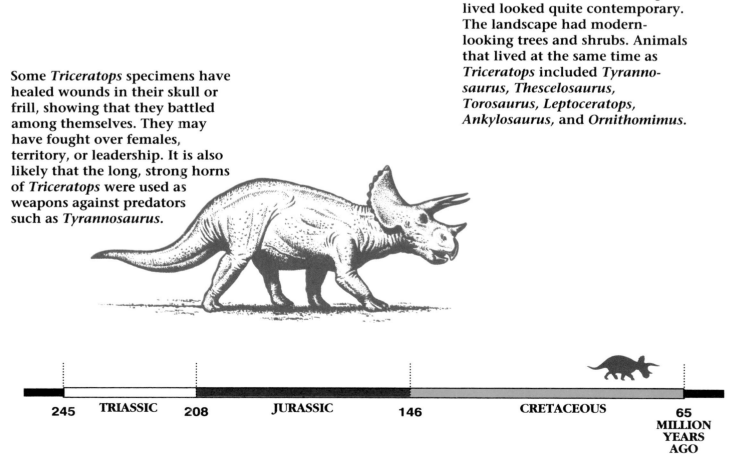

245	TRIASSIC	208	JURASSIC	146	CRETACEOUS	65 MILLION YEARS AGO

TYRANNOSAURUS
(tie-RAN-oh-SORE-us)

Tyrannosaurus rex—in Latin, it means "tyrant reptile king," and if ever there was a beast that could inspire both fear and respect, it was *Tyrannosaurus*. The largest known predator to have walked the earth, *Tyrannosaurus* measured about three car lengths, was as tall as a split-level house, and was topped by a head as big as an overstuffed chair!

Unlike some dinosaurs, *Tyrannosaurus* used all of its 60 teeth to eat meat, catching its dinner on the run using very powerful leg muscles. Its jaws (set with some teeth as big as bananas) were hinged so that the animal could swallow large chunks of meat. Another trait was *Tyrannosaurus*'s forward-looking eyes: Most meat-eating dinosaurs had eyes on the sides of their skulls, but the eyes of *Tyrannosaurus rex* looked ahead, perhaps giving it a special ability to judge depth and distance.

Tyrannosaurs attacking *Edmontosaurus regalis*. *Tyrannosaurus* is considered a born hunter. However, it was unlikely to have passed up a free meal if it came across a dead animal.

This king of reptiles was a fearsome opponent in battle and a tireless predator: With powerful thigh muscles and a strong tail, it was built for walking and running. In fact, *Tyrannosaurus* could hunt for an entire day, moving at the rate of four to five miles an hour. From time to time, it could also break into a 20-mile-an-hour run!

How do we know so much about *Tyrannosaurus rex?* A great deal can be learned from the seven incomplete skeletons and other assorted remains found over the years in the western United States and in one Canadian province.

The most recent finds come from Montana and South Dakota. In 1987, paleontologists discovered the most nearly complete specimen ever. The skull alone measures about five feet. The Montana dinosaur, which had two-clawed hands and robust arms, also is the first to be found with nearly complete forelimbs.

Other digs have unearthed *Tyrannosaurus* remains that indicate the reptile king averaged about 13 feet in height and something like 40 feet in length. However, at the University of California at Berkeley, there is a massive upper jawbone of *Tyrannosaurus* that suggests this particular dinosaur might have been nearly 16 feet tall and more than 50 feet long—and weighed more than the heaviest modern elephant!

Period:	
Late Cretaceous	
Order, Suborder, Family:	
Saurischia, Theropoda, Tyrannosauridae	
Location:	
North America (Canada, United States), Asia	
Length:	
40 feet (12 meters)	

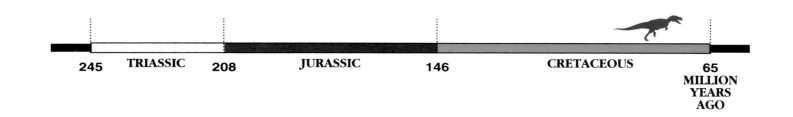

| 245 | TRIASSIC | 208 | JURASSIC | 146 | CRETACEOUS | 65 MILLION YEARS AGO |

Tyrannosaurus may have had small arms, but they were not weak: The robust front limbs, with their two-fingered hands, probably were used as hooks to hold struggling prey.

The largest known predator to have walked the earth, *Tyrannosaurus* measured about three car lengths, was as tall as a split-level house, and was topped by a head as big as an overstuffed chair!

A skeletal drawing of *Tyrannosaurus*. Considering the skull's enormous size, it was lightly built, with large openings and hollow bones. Some bones were not joined tightly, one reason *Tyrannosaurus* skulls are often found in pieces.

123

ANCHICERATOPS

(ANK-ee-SAIR-ah-tops)

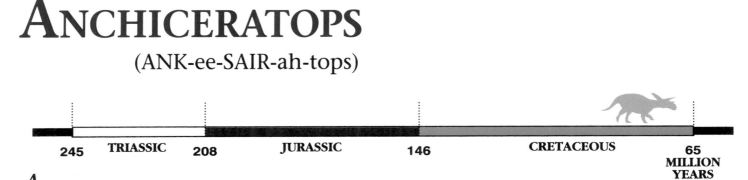

245	TRIASSIC	208	JURASSIC	146	CRETACEOUS	65 MILLION YEARS AGO

*A*nchiceratops was discovered along the Red Deer River in Alberta, Canada, in 1912 by an expedition from the American Museum of Natural History led by Barnum Brown. He found the back part of a skull that had a new kind of frill. Brown named it *Anchiceratops ornatus,* the "close-horned reptile."

In 1924, Charles Sternberg collected a nearly complete skull of *Anchiceratops,* which provided the information missing on the first skull. Sternberg noted that this skull was slightly different from the first one, so he designated it a new species, *Anchiceratops longirostris.* Paleontologists are not sure if there is more than one species of *Anchiceratops.* More specimens have since been found, with a nearly complete skeleton housed in the National Museum of Canada.

The most distinctive features of *Anchiceratops* are in its unusual neck frill. The frill is moderately long and rectangular with small, oval openings. The edge of the frill is thick just behind the brow horns. On the back of the frill are six large bony knobs that were expanded into short, triangular, backward-pointing spikes. Also on the frill are two short spikes that curve up and out. *Anchiceratops* had a short nasal horn, a very long nose, and two moderate-size brow horns. Its skeleton shows it had a very short tail, but otherwise it looked much like other ceratopsids.

Anchiceratops lived at the same time as its relative *Arrhinoceratops*. It was also closely related to *Torosaurus, Chasmosaurus, Pentaceratops,* and *Triceratops*.

Period:
Late Cretaceous
Order, Suborder, Family:
Ornithischia, Marginocephalia, Ceratopsidae
Location:
North America (Canada)
Length:
20 feet (6 meters)

Paleontologists are not sure if there is more than one species of *Anchiceratops*.

A herd of *Anchiceratops.* The "close-horned reptile," which looked much like other ceratopsids, was first discovered in Alberta, Canada, in 1912.

ANTARCTOSAURUS
(ant-ARK-toh-SORE-us)

With a thigh bone more than seven and a half feet long—longer than any other dinosaur thigh bone known—*Antarctosaurus* was a sauropod of spectacular proportions. Like other members of the Late Cretaceous family Titanosauridae, it had massive hips, tall rear legs, front legs nearly as tall as the rear, a long tail, and a long neck.

Antarctosaurus had a very small head, even for a sauropod. It was a plant-eater that had weak, peglike teeth. None of the preserved skeletons of *Antarctosaurus* are complete, so there is still much to learn about this animal.

Some paleontologists calculate that, based on the length of the limb bones, *Antarctosaurus* may have been larger than *Brachiosaurus brancai,* which is the largest dinosaur known from complete skeletons. But the two are not closely related; *Brachiosaurus* belongs to a different family and lived only in the Jurassic Period. *Antarctosaurus* was the largest dinosaur of the Late Cretaceous. It may have weighed between 80 to 100 tons and stood as tall as 15 feet at the shoulders.

Its name means "Antarctic reptile." It was found in South America, which in the Late Cretaceous was not far from Antarctica. Fragmentary remains from India have been found that may belong to *Antarctosaurus,* but scientists have not yet proved they belong to the same animal.

A relative, *Titanosaurus,* has been found in India and Argentina. This may mean that Gondwanaland still had land connections at that time.

Period:
Late Cretaceous
Order, Suborder, Family:
Saurischia, Sauropodomorpha, Titanosauridae
Location:
South America (Argentina, Brazil, Uruguay)
Length:
Estimated 80–100 feet (24–30 meters)

Antarctosaurus wichmannianus. The largest dinosaur of the Late Cretaceous, *Antarctosaurus* was found in South America, which long ago was much closer to the South Pole.

245	TRIASSIC	208	JURASSIC	146	CRETACEOUS	65 MILLION YEARS AGO

125

ARALOSAURUS

(ah-RAL-oh-SORE-us)

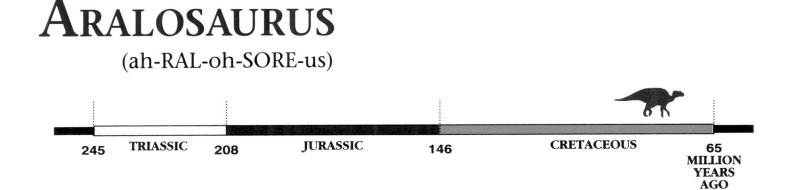

| 245 | TRIASSIC | 208 | JURASSIC | 146 | CRETACEOUS | 65 MILLION YEARS AGO |

The upper jaw of *Aralosaurus* was strongly built and relatively high. Like other duckbilled dinosaurs, it had many teeth for grinding plant food.

Period: Late Cretaceous
Order, Suborder, Family: Ornithischia, Ornithopoda, Hadrosauridae
Location: Asia (Kazakhstan)
Length: Unknown

*A*ralosaurus is from Kazakhstan in the former Soviet Union. It is known only from a skull that is missing the front of the snout and all of the lower jaw; no skeleton exists. This Late Cretaceous duckbilled dinosaur, whose name means "reptile from the Aral region," had a strongly hooked nasal arch in front of the eyes. This hook is better developed than in any other hadrosaurid, including the dinosaur's close relative *Gryposaurus*. *Hadrosaurus* may also have had an arched snout, but this part of the skeleton is not preserved. The hook of *Aralosaurus* was not only high, but also wide from side to side. The nostril openings beneath the arch were large.

The upper jaw of *Aralosaurus* was strongly built and relatively high. Like other duckbilled dinosaurs, it had many teeth for grinding plant food. *Aralosaurus* seems to have had as many as 30 tooth rows, with several teeth per row. The back of the skull was very high and wide. This suggests that there must have been long, massive muscles for chewing food.

Aralosaurus was first named and described in 1968. Since then, very little has been written about dinosaurs from this area of the world. Perhaps future efforts will tell us more about *Aralosaurus*, its anatomy, and its evolution.

Aralosaurus tuberiferus. The "reptile from the Aral region" is distinguished by its hooked nasal arch, strong upper jaw, and numerous grinding teeth.

ARRHINOCERATOPS
(ar-RINE-oh-SAIR-ah-tops)

Arrhinoceratops is a rare ceratopsian known from only one skull that lacks a lower jaw. This single specimen was found in 1923 along the Red Deer River in Alberta, Canada, by an expedition from the University of Toronto. William A. Parks named this dinosaur *Arrhinoceratops brachyops,* meaning "no nose-horn face," in 1925. The name is incorrect, however: *Arrhinoceratops* did have a short, blunt nasal horn.

Arrhinoceratops had a short, deep, wide face with large nostrils. Its two brow horns were moderately long, very pointed, and curved forward. The neck frill was broad with small, oval openings, and it was rounded. Much of the skull of *Arrhinoceratops* was slightly crushed and distorted, making it difficult to understand the pattern of the skull bones. Paleontologists do not know what the rest of its body looked like.

Arrhinoceratops lived at the same time as its close relative *Anchiceratops.* It was also closely related to *Torosaurus, Chasmosaurus, Pentaceratops,* and *Triceratops.*

Period:
Late Cretaceous
Order, Suborder, Family:
Ornithischia, Marginocephalia, Ceratopsidae
Location:
North America (Canada)
Length:
20 feet (6 meters)

Despite its name, which means "no nose-horn face," *Arrhinoceratops brachyops* did have a short nasal horn, as can be seen from this skull lacking a lower jaw—the only fossil available of this rare dinosaur.

Arrhinoceratops brachyops. A close relative of *Torosaurus* and *Chasmosaurus,* the dinosaur had a short, wide face and moderately long brow horns.

245	TRIASSIC	208	JURASSIC	146	CRETACEOUS	65 MILLION YEARS AGO

AUBLYSODON

(oh-BLISS-oh-don)

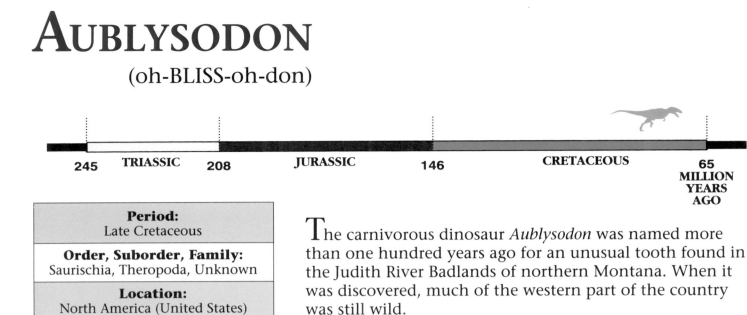

245	**TRIASSIC**	208	**JURASSIC** 146	**CRETACEOUS** 65 **MILLION YEARS AGO**

Period:	
Late Cretaceous	
Order, Suborder, Family:	
Saurischia, Theropoda, Unknown	
Location:	
North America (United States)	
Length:	
Unknown	

The carnivorous dinosaur *Aublysodon* was named more than one hundred years ago for an unusual tooth found in the Judith River Badlands of northern Montana. When it was discovered, much of the western part of the country was still wild.

Since then, many paleontologists have offered different opinions about the tooth. Most have accepted the genus *Aublysodon,* although reluctantly. The reason is that, during the early days of paleontology, many names of dinosaurs were given to very scant material. Some scientists have argued that the single tooth, although unusual, might belong to a dinosaur already known.

The mystery may have been finally solved when a partial skull was found recently in Montana. The skull shows a long, low snout and an unusual step in the lower jaw. Unfortunately, so little is known of this animal that we do not know its size or weight. We do know that *Aublysodon* was widespread; its unusual teeth have been found in many states.

Though little is known about its appearance or habits, *Aublysodon* clearly was a meat-eater—and a widespread one: Its distinctive teeth have been found in a number of states.

128

AVACERATOPS
(AYV-ah-SAIR-ah-tops)

Avaceratops lammersi was a small ceratopsid known from a single skeleton found in the Judith River Formation of Montana in 1981. The bones of *Avaceratops* were scattered in a fossil stream. After death, the animal was probably washed down the stream and buried in a sand bar.

Avaceratops was only about seven and a half feet long, and some paleontologists believe it may have been a juvenile. Others think that it was an adult and that *Avaceratops* may have been a small ceratopsid. Although only slightly larger than *Protoceratops*, *Avaceratops* appears to have had a moderately heavy build, like its larger ceratopsid relatives. Unfortunately, many pieces of the skeleton are missing.

Avaceratops had a short, deep snout and a thick, powerful lower jaw. The neck frill appears to have been solid. Since the top of the skull is missing, it is not known what kind of horns it had—or if it had any. Most of its closest relatives, including *Centrosaurus*, *Styracosaurus*, and *Brachyceratops*, had a large nasal horn and no brow horns, so *Avaceratops* may have looked like them.

Because of its small size, *Avaceratops* probably ate low vegetation, which would have been mostly flowering plants.

Many other animals lived alongside *Avaceratops*, including hadrosaurs, pachycephalosaurs, ankylosaurs, hypsilophodontids, dromaeosaurs, crocodilians, turtles, and small mammals.

Period:
Late Cretaceous
Order, Suborder, Family:
Ornithischia, Marginocephalia, Ceratopsidae
Location:
North America (United States)
Length:
7¹/₂ feet (2.3 meters)

Because of its small size, *Avaceratops* probably ate low vegetation, which would have been mostly flowering plants.

Avaceratops lammersi, named by Peter Dodson for Ava Cole, whose husband found the only known skeleton on the Lammers family's Montana ranch in 1981.

245	**TRIASSIC**	208	**JURASSIC**	146	**CRETACEOUS**	65 **MILLION YEARS AGO**

AVIMIMUS

(AYV-ee-MIME-us)

*A*vimimus ("bird mimic") was a small, lightly built theropod from the Late Cretaceous Djadokhta Formation of Mongolia. The original specimens were collected by Russian scientists, and several partial skeletons and skulls have since been found. The most nearly complete skeleton is a partial skull, bones from the neck and back, a partial arm, most of both back legs and feet, and part of the pelvis.

The skull was short and birdlike; the dinosaur was toothless and had a beak and a long neck. The bones that surrounded and protected the brain were big, showing *Avimimus* may have had a large brain. The hole in the back of the skull where the spinal cord passed through was large for a dinosaur this size. But the occipital condyle—the bone at the back of the skull connecting to the neck—is small; this shows the skull was light for its size.

The largest of the two forearm bones had a ridge similar to rows of bumps on similar bones in modern birds. These bumps are where the quills of large flight feathers attach. Paleontologist Seriozha Kurzanov has claimed that this is evidence that *Avimimus* had feathers. The original skeleton was missing a tail, which led Kurzanov to believe that *Avimimus* did not have a tail. Since then, however, tail bones have been found with other fossils. The back legs of *Avimimus* were long in relation to its thigh bones, showing it was a fast runner.

Kurzanov thinks this animal fed mainly on insects. Others have suggested it ate plants. More information is needed to determine the exact nature of this unusual dinosaur, its habits, and its relationship to other dinosaurs.

Avimimus portentosus.
Paleontologist Seriozha Kurzanov believes that the many birdlike features in the skull and arm of *Avimimus* show it was capable of weak flight. This would have made it a feathered, flying theropod dinosaur that developed separately from true birds. Most scientists, however, disagree.

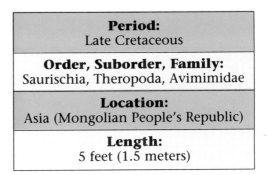

Period:
Late Cretaceous

Order, Suborder, Family:
Saurischia, Theropoda, Avimimidae

Location:
Asia (Mongolian People's Republic)

Length:
5 feet (1.5 meters)

The bones that surrounded and protected the brain were big, showing *Avimimus* may have had a large brain. The hole in the back of the skull where the spinal cord passed through was large for a dinosaur this size.

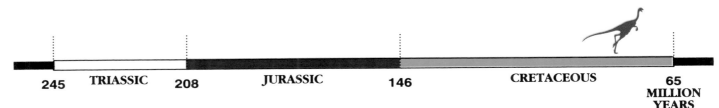

| | 245 | TRIASSIC | 208 | JURASSIC | 146 | CRETACEOUS | 65 MILLION YEARS AGO |

BACTROSAURUS
(BACK-troh-SORE-us)

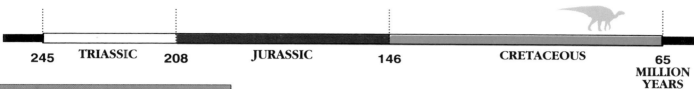

| 245 | TRIASSIC | 208 | JURASSIC | 146 | CRETACEOUS | 65 MILLION YEARS AGO |

Period:
Late Cretaceous
Order, Suborder, Family:
Ornithischia, Ornithopoda, Hadrosauridae
Location:
Asia (People's Republic of China)
Length:
20 feet (6 meters)

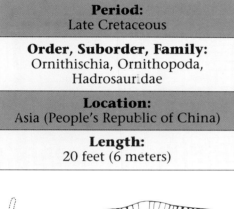

A skeletal drawing of *Bactrosaurus*. The dinosaur is known from numerous bones, but no complete skeleton exists. It is still one of the best understood early duckbilled dinosaurs.

Very little is known about the earliest duckbilled dinosaurs (hadrosaurids). The best known of the earliest are *Gilmoreosaurus* and *Bactrosaurus* from the Inner Gobi Desert of the People's Republic of China. The rocks from which these two hadrosaurids were found are dated as early Late Cretaceous, but they may be somewhat later than that. Further research in the field and in the laboratory now being done in China, Canada, and the United States will provide more information.

Bactrosaurus ("reptile from Bactria") is known from several skull and skeletal pieces, but not a complete skeleton. Many of these bones are from juveniles and perhaps hatchlings, while others are from small adults. The pelvis, limbs, and most of the important parts of the head, including jaws and teeth, are known for *Bactrosaurus*.

For a long time, it was thought that *Bactrosaurus* was different from all other lambeosaurine hadrosaurids because it did not have a crest. That would have made it quite unusual. But recent work on both *Bactrosaurus* and *Gilmoreosaurus* has finally sorted out what bones go with which animal. This helped identify parts of the base of a crest—similar to the kind found on the dinosaur's close relative *Parasaurolophus*.

Bactrosaurus johnsoni, from early Late Cretaceous rocks in the Inner Gobi Desert. Paleontologists recently found evidence that the dinosaur had a crest, like other lambeosaurine hadrosaurids.

131

BAGACERATOPS

(BAG-ah-SAIR-ah-tops)

Bagaceratops rozhdestvenskyi was a small protoceratopsian with a big name: "Baga" is the Mongolian word for "small," "ceratops" means "horned face," and the species name is in honor of the Russian paleontologist A. K. Rozhdestvensky. *Bagaceratops* was discovered in the Gobi Desert of Mongolia in the early 1970s by the Joint Polish-Mongolian Paleontological Expeditions. Specimens of *Bagaceratops* are at the Paleobiological Institute in Warsaw, Poland.

Bagaceratops was one of the smallest and most primitive of the known protoceratopsid dinosaurs. Very little is known about the skeleton of *Bagaceratops,* as only a few fragments were found. It was probably similar to that of *Protoceratops.* However, paleontologists did find partial to nearly complete skulls of both juveniles and adults. One of the tiny juvenile skulls is only about the size of a golf ball.

The skull of *Bagaceratops* had a short, low snout that was topped by a nasal bump, rather than a horn. The neck frill was triangular and very short, and it had no openings. The dinosaur had only ten grinding teeth in each jaw and no teeth in its beak.

Other animals that lived with *Bagaceratops* include the carnivorous dinosaurs *Velociraptor* and *Oviraptor,* as well as ankylosaurs, lizards, and small mammals.

> *Bagaceratops* was one of the smallest and most primitive of the known protoceratopsid dinosaurs. . . . One of the tiny juvenile skulls is only about the size of a golf ball.

Period:
Late Cretaceous

Order, Suborder, Family:
Ornithischia, Marginocephalia, Protoceratopsidae

Location:
Asia (Mongolian People's Republic)

Length:
5 feet (1.5 meters)

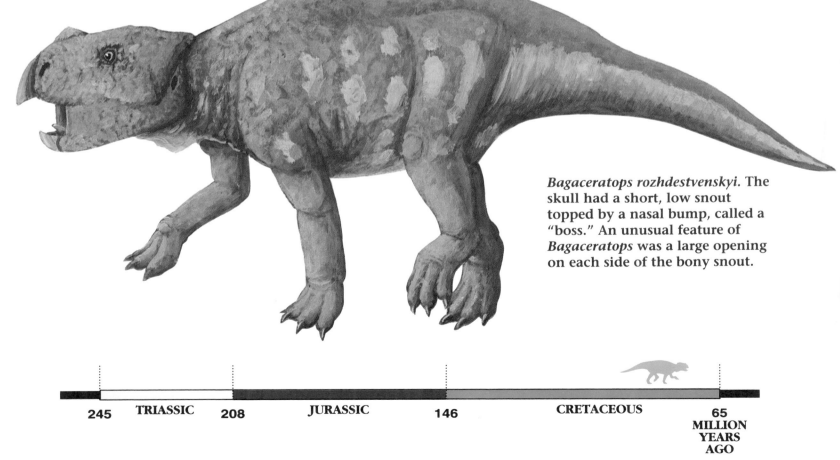

Bagaceratops rozhdestvenskyi. The skull had a short, low snout topped by a nasal bump, called a "boss." An unusual feature of *Bagaceratops* was a large opening on each side of the bony snout.

245	TRIASSIC	208	JURASSIC	146	CRETACEOUS	65 MILLION YEARS AGO

BRACHYCERATOPS
(BRAK-ee-SAIR-ah-tops)

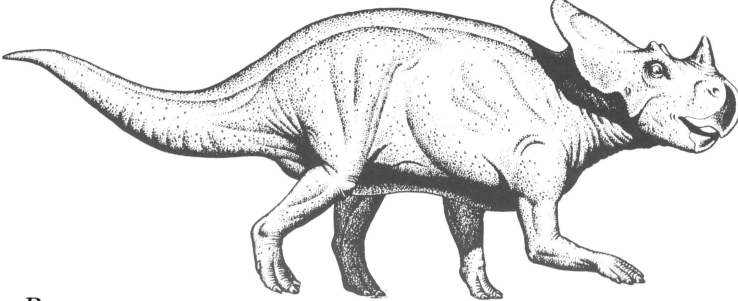

*B*rachyceratops montanensis was found in 1913 by paleontologist Charles W. Gilmore on the Blackfoot Indian Reservation in Montana. He found parts of at least five animals of the same size, all jumbled together. In this mix of *Brachyceratops* bones was a single incomplete and unattached skull. These fossils are now at the Smithsonian Institution in Washington, D.C. Although a few other bones of *Brachyceratops* have since been found, it remains a rare ceratopsian.

Because *Brachyceratops* was so small and the skull Gilmore found was in pieces, many paleontologists believe that these *Brachyceratops* specimens were juveniles. Some paleontologists think that *Brachyceratops* might be a young *Monoclonius,* but this has not been proven. It is probable that *Brachyceratops* is a separate genus.

Gilmore's discovery of these five small animals together is very unusual. If they were juveniles, as seems likely, they may have been nest mates that remained together after hatching.

Brachyceratops had a low, thick nasal horn; small bumps over the eyes (but no real brow horns); and a moderate-size neck frill. Since some pieces of the frill are missing, it is not known if it had any openings. Since the fossil remains may have been those of a juvenile, we do not know how large the dinosaur was as an adult.

Brachyceratops was a centrosaurine ceratopsid; its closest relatives were *Avaceratops, Centrosaurus, Monoclonius, Styracosaurus,* and *Pachyrhinosaurus.*

Questions surround *Brachyceratops montanensis*—was this rare ceratopsian actually a juvenile *Monoclonius,* or was it a separate genus? And if a juvenile, how big was the adult?

Period:
Late Cretaceous
Order, Suborder, Family:
Ornithischia, Marginocephalia, Ceratopsidae
Location:
North America (Canada, United States)
Length:
5 feet (1.5 meters)

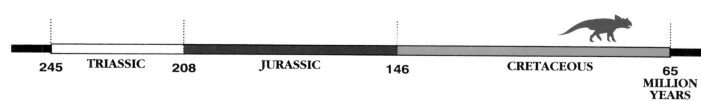

| 245 | TRIASSIC | 208 | JURASSIC | 146 | CRETACEOUS | 65 MILLION YEARS AGO |

BRACHYLOPHOSAURUS

(BRAK-ee-LOAF-oh-SORE-us)

Most crests of hadrosaurids like *Brachylophosaurus*, from simple arches and bumps to elaborate hollow crests, may be related to how the animal behaved. They might have been used in contests of strength to defend territory or females, or for display to scare enemies away.

Brachylophosaurus ("short-ridged reptile") was a duckbilled dinosaur from the Late Cretaceous. This hadrosaurid was discovered and named by Charles M. Sternberg of Ottawa, Canada, in 1953. In some ways, it was similar to other hadrosaurids—it had many functional and replacement teeth and a flaring, ducklike snout. But *Brachylophosaurus* had a broad, flat, shieldlike crest on top of its head, directly above its eyes. This bony crest is unlike that of any other hadrosaurid.

The upper beak of *Brachylophosaurus* was larger and broader than in any other duckbilled dinosaur from the same time. The upper beak also turned down where the horny covering was attached. Like other large ornithopods, the dinosaur used this horny beak to nip foliage from low-standing boughs and shrubs.

Brachylophosaurus had many teeth behind its beak for chewing plant food. Like other hadrosaurids (and many ornithopods), chewing was done by an unusual movement of the upper jaws with the large jaw muscles at the back of the skull. When the upper and lower teeth touched, the upper jaw moved slightly to the side to allow the teeth to move past one another. In this way, these hadrosaurids were able to chew with a sideways motion, somewhat similar to cows and horses. Muscular cheeks kept food from spilling out of the sides of the mouth.

Brachylophosaurus had very long arms; the reason for this is not known. They were not as long in the dinosaur's relatives *Maiasaura* and *Prosaurolophus*.

Period:
Late Cretaceous

Order, Suborder, Family:
Ornithischia, Ornithopoda, Hadrosauridae

Location:
North America (Canada, United States)

Length:
30 feet (9 meters)

. . . Brachylophosaurus **had a broad, flat, shieldlike crest . . . unlike that of any other hadrosaurid.**

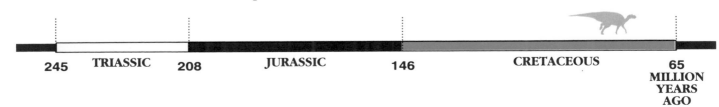

| 245 | TRIASSIC | 208 | JURASSIC | 146 | CRETACEOUS | 65 MILLION YEARS AGO |

CENTROSAURUS
(SEN-troh-SORE-us)

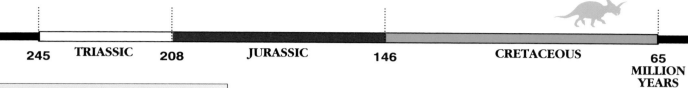

245	TRIASSIC 208	JURASSIC 146	CRETACEOUS 65 MILLION YEARS AGO

Period:	Late Cretaceous
Order, Suborder, Family:	Ornithischia, Marginocephalia, Ceratopsidae
Location:	North America (Canada)
Length:	17 feet (5 meters)

Centrosaurus, which means "sharp-point reptile," was named by Lawrence Lambe in 1902 from specimens found along the Red Deer River in Alberta, Canada. A number of complete skulls and skeletons have since been discovered. An entire *Centrosaurus* herd, ranging in size from juveniles to older adults, was found in a fossil river bed in Dinosaur Provincial Park in southern Alberta. Paleontologists think this herd may have drowned while crossing a flood-swollen river.

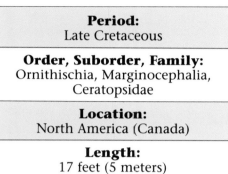

A skull of *Centrosaurus apertus.* Sometimes confused with *Monoclonius* because of similar skull features, *Centrosaurus* has recently been shown to be a separate genus.

Centrosaurus resembled its close relative *Monoclonius. Centrosaurus* had a long, pointed nasal horn; small bumps over its eyes instead of brow horns; and a short, rounded neck frill with moderately large openings. *Monoclonius* and *Centrosaurus* lived at the same time and in the same places. Because they looked alike and because of the poor quality of most *Monoclonius* specimens, *Centrosaurus* and *Monoclonius* have been confused with each other or even considered the same dinosaur.

Recent studies have confirmed that they are separate dinosaurs. Part of the proof is the unique paired hornlike growths at the back of the frill of *Centrosaurus*. One pair of growths were long, grooved, banana-shaped tongues of bone that curved forward and down over the openings in the neck frill. A second pair grew backward and curved toward each other.

Other close relatives of *Centrosaurus* included *Styracosaurus, Pachyrhinosaurus, Brachyceratops,* and *Avaceratops.*

A centrosaur herd fording a swollen river. Just such a scene might have taken place in what is now Canada's Dinosaur Provincial Park, where fossils from an entire *Centrosaurus* herd were found in an ancient river bed—presumably, the animals had drowned while trying to cross.

135

CONCHORAPTOR

(CONK-oh-RAP-ter)

Period:
Late Cretaceous
Order, Suborder, Family:
Saurischia, Theropoda, Oviraptoridae
Location:
Asia (Mongolian People's Republic)
Length:
4–6 feet (1–2 meters)

The oviraptorids were peculiar theropods. Smallish animals that walked on two legs, they had strong beaks and may have fed on mollusks by crushing their shells to get the soft meat inside. For many years, only one oviraptorid, *Oviraptor philoceratops*, was known. Some paleontologists classified it as an ornithomimid because its skull was toothless.

More oviraptorid specimens were discovered in the 1970s. These have been examined, and descriptions were published of two new animals, *Oviraptor mongoliensis* and *Conchoraptor gracilis*. The name of the latter species means "slender conch-stealer." Most scientists now keep the oviraptorids in their own family, separate from the "ostrich dinosaurs."

Conchoraptor was smaller than its relative *Oviraptor*. The head of *Oviraptor* was decorated with bony crests, but *Conchoraptor* had no decoration. At first, it was thought that *Conchoraptor* was a juvenile *Oviraptor* and that the cranial crest developed at the beginning of sexual maturity. Further study of more skeletons—especially the hands—showed that *Conchoraptor* was a different genus. Its hands seem to be a transitional form, or "missing link," between *Oviraptor* and the oviraptoridlike small theropod *Ingenia*.

Conchoraptor was smaller than its relative *Oviraptor*. . . . Its hands seem to be a transitional form, or "missing link," between *Oviraptor* and the oviraptoridlike small theropod *Ingenia*.

Conchoraptor gracilis. Discovered in Mongolia's Nemegt Formation, the dinosaur and its relative *Oviraptor* are considered distinct from the ornithomimids, despite their similarly toothless skulls.

| 245 | TRIASSIC | 208 | JURASSIC | 146 | CRETACEOUS | 65 MILLION YEARS AGO |

CORYTHOSAURUS
(coh-RITH-oh-SORE-us)

Originally found and named by Barnum Brown of the American Museum of Natural History, *Corythosaurus* is one of the best known of all dinosaurs.

It had a hollow crest on top of its head above its eyes; this is often considered the most striking aspect of the animal. Its lambeosaurine relatives *Lambeosaurus, Hypacrosaurus,* and *Parasaurolophus* also had crests. The dinosaur's crest, which had different shapes and sizes in males and females and developed only in mature adults, contained the nasal cavity. When *Corythosaurus* breathed, air entered the nostrils. It then went up into the chambers of the crest, into large side pockets, and then to a common chamber in the center of the crest. From there, air traveled down to the back of the throat and into the wind pipe to the lungs.

Returning air took the opposite course. So did sounds that the animal may have made. In fact, it looks as if the crest would have made an excellent resonator, like a wind or brass instrument. The dinosaur would have made very low sounds. Scientists think that *Corythosaurus* would have used these low trumpeting notes to make calls to each other or to offspring. But the young, like young alligators today, would have chirped more highly-pitched sounds to their parents. Perhaps these calls would have been about food or water or a predator in the area.

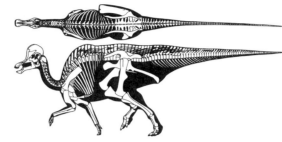

Skeletal drawings of *Corythosaurus*. Many skeletons of this animal have been found, some with complete skulls. Others were partial "mummies" with skin impressions and impressions of a horny bill. Some still had middle ear elements in place.

Period:
Late Cretaceous
Order, Suborder, Family:
Ornithischia, Ornithopoda, Hadrosauridae
Location:
North America (Canada, United States)
Length:
33 feet (10 meters)

Corythosaurus. The "corinthian helmet reptile," it was one of the most abundant duckbilled dinosaurs from the Late Cretaceous of western North America.

| 245 | TRIASSIC | 208 | JURASSIC | 146 | CRETACEOUS | 65 MILLION YEARS AGO |

DASPLETOSAURUS

(das-PLEE-toh-SORE-us)

With its massive head and large teeth, *Daspletosaurus* was truly a master of its world. It got its name because of its ferociousness; the name means "frightful reptile." It had a huge body balanced upon two powerful back legs. The three-toed taloned feet (much like those of a modern bird) probably held the prey down while *Daspletosaurus* ate. As in all tyrannosaurids, the front limbs were short and had only two fingers each.

Daspletosaurus lived at the same time as *Albertosaurus*. How two large, meat-eating dinosaurs could have lived side by side is a mystery. Perhaps it was much like the lion and the cheetah in East Africa today. These two cats have different methods of hunting, and they prey on different animals. The lion relies on stealth to get close to its prey (gnu, zebra, etc.), while the slender, fast cheetah runs down its prey. It is possible that *Daspletosaurus,* with its massive head and body, might have stalked the ceratopsians, while the slenderer, quicker *Albertosaurus* might have run down hadrosaurs. *Daspletosaurus* and ceratopsians are less common than *Albertosaurus* and hadrosaurs in the badlands of Canada's Alberta Province.

Period:
Late Cretaceous
Order, Suborder, Family:
Saurischia, Theropoda, Tyrannosauridae
Location:
North America (Canada)
Length:
30 feet (9 meters)

A skeleton of *Daspletosaurus torosus* on display at Ottawa's Canadian Museum of Nature.

As in *Tyrannosaurus,* the short, two-fingered hands of *Daspletosaurus* contrasted the powerful hind legs with their three-toed feet. Perhaps *Daspletosaurus* stalked its prey rather than running it down.

With its massive head and large teeth, *Daspletosaurus* was truly a master of its world. It got its name because of its ferociousness; the name means "frightful reptile."

245	TRIASSIC	208	JURASSIC	146	CRETACEOUS	65 MILLION YEARS AGO

DROMAEOSAURUS
(DROH-may-oh-SORE-us)

Period:
Late Cretaceous
Order, Suborder, Family:
Saurischia, Theropoda, Dromaeosauridae
Location:
North America (Canada, United States)
Length:
6 feet (1.8 meters)

In 1914, Barnum Brown of the American Museum of Natural History collected a nine-inch-long skull and some foot bones from the Judith River Formation in Alberta, Canada. The skeletal material was named *Dromaeosaurus,* which means "running reptile." After that, the genus was often referred to in various publications, even though very little was known about it.

In 1969, the skull and foot bones were described again, and some important similarities between *Dromaeosaurus* and the dinosaur *Deinonychus* were noticed. Both were small theropods with large skulls. Their skulls were similar, and both had large, sicklelike claws on the second toe of the foot. The two dinosaurs are now placed in the same family, Dromaeosauridae. Other members of this family include *Velociraptor* and *Hulsanpes* from Mongolia.

Not much is known about *Dromaeosaurus* because the only specimen is incomplete. Teeth that may belong to it have been found in several western states and Alberta, but they only tell us that the animal lived in these areas. This was a rare theropod—at best, it was rarely preserved. Scientists are hoping better specimens will be found.

A *Dromaeosaurus* skull. The "running reptile" was first found in 1914 in Canada, but teeth from the dinosaur may have since been found throughout the American West.

In 1969, the skull and foot bones were described again, and some important similarities between *Dromaeosaurus* and the dinosaur *Deinonychus* were noticed.

Dromaeosaurus albertensis. A rare predator, it has a number of features in common with *Deinonychus* and another speedy meat-eater, *Velociraptor.*

245	TRIASSIC	208	JURASSIC	146	CRETACEOUS	65 MILLION YEARS AGO

DROMICEIOMIMUS

(droh-MEE-see-oh-MY-mus)

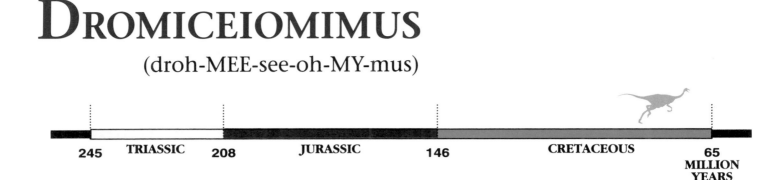

| 245 | **TRIASSIC** | 208 | **JURASSIC** | 146 | **CRETACEOUS** | 65 **MILLION YEARS AGO** |

Dromiceiomimus ("emu mimic") has been found both in the Late Cretaceous Horseshoe Canyon and in the Judith River Formation of Alberta, Canada. It is very similar to *Struthiomimus* and *Ornithomimus,* but had much larger eyes and longer, more slender arms. Also, some of the hip bones were positioned differently.

As in all ornithomimids, the brain of *Dromiceiomimus* was quite large. This does not mean, however, that the animal was intelligent. For example, the ostrich and emu also have relatively large brains; they are not very smart, but they have very good vision. The enlarged portions of the dinosaur's brain probably were to coordinate body and limb actions. *Dromiceiomimus* was probably omnivorous; it is likely that it ate fruits, large insects, and small lizards and mammals.

Dromiceiomimus had birdlike jaws with no teeth. Another birdlike feature was the dinosaur's hollow bones. With its long legs, it ran swiftly.

| **Period:** |
| Late Cretaceous |
| **Order, Suborder, Family:** |
| Saurischia, Theropoda, Ornithomimidae |
| **Location:** |
| North America (Canada) |
| **Length:** |
| 12 feet (3.5 meters) |

Dromiceiomimus, a swift, birdlike predator, was toothless. Nevertheless, it ate insects, small lizards, and mammals—and perhaps fruits, as well.

As in all ornithomimids, the brain of *Dromiceiomimus* was quite large. This does not mean, however, that the animal was intelligent. . . . The enlarged portions of the dinosaur's brain probably were to coordinate body and limb actions.

EDMONTONIA
(ED-mon-TONE-ee-uh)

Period:	Late Cretaceous
Order, Suborder, Family:	Ornithischia, Thyreophora, Nodosauridae
Location:	North America (Canada, United States)
Length:	22 feet (6.6 meters)

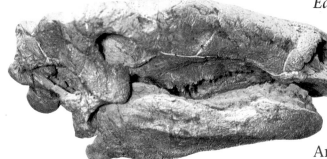

An *Edmontonia longiceps* skull. When viewed from above, the dinosaur's skull was pear shaped.

Edmontonia would not have made an easy meal for a hungry tyrannosaur. It had a heavily armored body and large, forward-pointing shoulder spines. We know what *Edmontonia* looked like because two specimens were found with their armor and spikes preserved in the position they had in life. The bodies of these specimens may have dried out because of a drought and then been quickly covered by sediment when the rainy season began. Evidence for these changes in the climate is found in the growth rings of fossil wood.

Edmontonia walked on four legs and was a plant-eater. It had a pear-shaped skull when viewed from the top. The neck and part of the back were protected by large, flat, ridged plates. Smaller ridged plates covered the back, hips, and tail. Spines and large spikes along its sides would have made the animal look short and wide when viewed from the front—and more menacing to an enemy.

An *Albertosaurus* was not the only animal it needed to protect itself from. A male *Edmontonia* probably fought with other males for territory and mates. The larger males may have used their shoulder spines for shoving contests. The spines of *Edmontonia* would have been dangerous to rival males or to *Albertosaurus* or *Daspletosaurus,* if either predator got too close.

Two *Edmontonia rugosidens* preparing to do battle. Spikes and spines not only made this plant-eater *look* fearsome: They could be painful, if not downright dangerous, to a predator or rival.

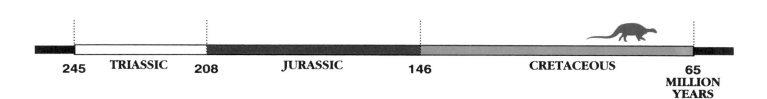

245	TRIASSIC	208	JURASSIC	146	CRETACEOUS	65 MILLION YEARS AGO

141

ERLIKOSAURUS

(AIR-lick-oh-SORE-us)

Period:
Late Cretaceous
Order, Suborder, Family:
Saurischia, Segnosauria, Segnosauridae
Location:
Asia (Mongolian People's Republic)
Length:
20 feet (6 meters)

The front of the snout was toothless, and the skull bones show that *Erlikosaurus*—and possibly other segnosaurians—had a horny beak for cropping plants.

In 1980, Altangerel Perle named *Erlikosaurus andrewsi* after the demon Erlik from Mongolian mythology and paleontologist Roy Chapman Andrews. It was closely related to *Segnosaurus*. *Erlikosaurus* is the only segnosaurian that was found with a skull, making it a very important member of the family. Unfortunately, little of the skeleton was discovered.

The bones of its foot were more slender than those of *Segnosaurus*. *Erlikosaurus* may have been a smaller, more lightly built animal. The front of the snout was toothless, and the skull bones show that *Erlikosaurus*—and possibly other segnosaurians—had a horny beak for cropping plants.

The teeth of *Erlikosaurus* show it was a plant-eater. How it used the large, thin, slender claws on its feet is a mystery; perhaps they were useful when the animal waded across rivers.

The segnosaurian *Enigmosaurus mongoliensis* is known only from part of a pelvis; it was found in the same formation as *Erlikosaurus*. Because *Segnosaurus* was closely related to *Erlikosaurus*, scientists think that the pelvis of *Erlikosaurus* was similar to that of *Segnosaurus*. The pelvis of *Enigmosaurus* was very different from the pelvis of *Segnosaurus*, so paleontologist Rinchen Barsbold did not think that it belonged to *Erlikosaurus*.

Mongolian dinosaurs are full of surprises, however, and scientists may find that *Erlikosaurus* and *Enigmosaurus* are the same animal.

Erlikosaurus andrewsi. Equipped with a horny beak and slender claws, *Erlikosaurus* is closely related to *Segnosaurus*, though a smaller and more delicate animal.

| 245 | TRIASSIC | 208 | JURASSIC | 146 | CRETACEOUS | 65 MILLION YEARS AGO |

EUOPLOCEPHALUS

(YOO-op-loh-SEF-uh-lus)

| 245 | TRIASSIC | 208 | JURASSIC | 146 | CRETACEOUS | 65 MILLION YEARS AGO |

Euoplocephalus lived at the same time and in the same areas as *Edmontonia*. *Euoplocephalus* roamed the forests, cropping low plants with its broad beak. It probably ate any type of plant it came across. It used its small teeth to chew.

Scientists do not know exactly what *Euoplocephalus* looked like in life because no specimen has been found with the armor preserved in its natural position. We do know that the head and body were well armored and that the armor was joined to the head. Even the eyes were protected with bone eyelids. This is how the dinosaur got its name, which means "well-armored head."

Though scientists do not know the exact arrangement, they do know the dinosaur had bands of armor on its back with large spikes protecting its neck and the base of its tail. It probably had smaller spikes all along its back. As was true of all ankylosaurids, *Euoplocephalus* had a tail club that it used as a weapon.

In a world with dangerous predators, a large, slow-moving animal needed ways to protect itself. All ankylosaurs had body armor. But *Euoplocephalus* may also have had a good sense of smell. The air passage in the nostrils was looped, so many sensory nerves for smell were probably present. It may have picked up the scent of a predator before it was seen. These looped passages also may have been used to make a trumpeting noise so that *Euoplocephalus* could communicate with other members of its species.

Euoplocephalus was closely related to other ankylosaurs, including *Talarurus, Saichania,* and *Ankylosaurus.* Some ankylosaurs survived until the very end of the Cretaceous Period.

| **Period:** |
| Late Cretaceous |
| **Order, Suborder, Family:** |
| Ornithischia, Thyreophora, Ankylosauridae |
| **Location:** |
| North America (Canada, United States) |
| **Length:** |
| 20 feet (6 meters) |

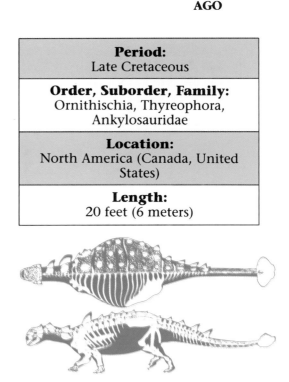

Skeletal drawings of *Euoplocephalus.* Well armored, the dinosaur had a stout tail club that it used to defend itself. It was a plant-eater with small teeth.

Euoplocephalus tutus. Member of a family of dinosaurs, some of which survived to the end of the Cretaceous, *Euoplocephalus* was endowed with excellent protective armor and possibly a keen sense of smell.

143

GALLIMIMUS
(GAL-lee-MIME-us)

Gallimimus ("chicken mimic") was the largest known member of the ornithomimids, or "ostrich dinosaurs." It has been found only in the Late Cretaceous Nemegt Formation of Mongolia. Its form and lifestyle were not very different from its North American relatives *Struthiomimus* and *Ornithomimus*. However, its back limbs were shorter, relative to its size; its toothless skull was much longer; and its hands were smaller when

compared to other members of the ornithomimid family. Several juveniles of this species also have been found.

Gallimimus could not use its hands to grasp, so it may have used them to dig in the ground for food. It is also possible *Gallimimus* fed on eggs. It probably ate mostly plants and may have also eaten insects if it could catch them. Like all theropods, it walked on its hind legs.

The largest of the "ostrich dinosaur" family, *Gallimimus* still had shorter hind limbs, compared to its size, than its relatives *Struthiomimus* and *Ornithomimus*. Its hands were also smaller and not capable of grasping.

Period:
Late Cretaceous
Order, Suborder, Family:
Saurischia, Theropoda, Ornithomimidae
Location:
Asia (Mongolian People's Republic)
Length:
17 feet (5 meters)

A *Gallimimus bullatus* skull. Equipped with a toothless beak, *Gallimimus* may have fed on plants, insects, and eggs. Its name means "chicken mimic."

A skeletal drawing of *Gallimimus*. Like all theropods, it was bipedal—and probably used its arms and hands for digging.

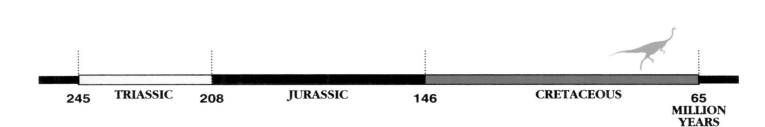

245	**TRIASSIC**	**208**	**JURASSIC**
	146	**CRETACEOUS**	**65 MILLION YEARS AGO**

GARUDIMIMUS
(gah-ROO-dee-MIME-us)

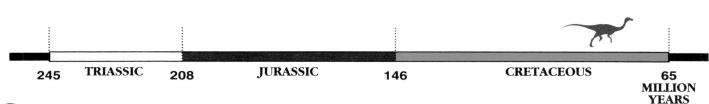

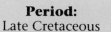

| 245 | TRIASSIC | 208 | JURASSIC | 146 | CRETACEOUS | 65 MILLION YEARS AGO |

Garudimimus was a primitive ornithomimid. Although its toothless skull looked like those of the other ornithomimids, it had a short, hornlike crest on top of its muzzle. The pelvis and feet of *Garudimimus* were not as adapted for running as those of its later relatives. Its foot still had a small first toe; this toe was gone in later ornithomimids. *Garudimimus* was placed in its own family, Garudimimidae, but many paleontologists think *Garudimimus* should remain in the family Ornithomimidae.

Despite the differences from its later relatives, *Garudimimus* still had many similarities. It walked on two legs and was a plant-eater. Its hands were not able to grasp, so it may have used them to dig for food.

In 1984, paleontologists described an earlier, even more primitive Mongolian ornithomimid, *Harpymimus okladnikovi*. All other ornithomimids were toothless, but *Harpymimus* had several very small teeth at the front of its snout, which the later ornithomimids did not have. Although they placed the animal in a new family, Harpymimidae, it too probably belongs in the family Ornithomimidae.

Because of these two primitive ornithomimids, Mongolia has become an important area for studying the origins of this theropod family.

Period:
Late Cretaceous
Order, Suborder, Family:
Saurischia, Theropoda, Ornithomimidae
Location:
Asia (Mongolian People's Republic)
Length:
12 feet (3.6 meters)

The pelvis and feet of *Garudimimus* were not as adapted for running as those of its later relatives. Its foot still had a small first toe; this toe was gone in later ornithomimids.

Garudimimus brevipes **was one of many dinosaurs found by the Joint Soviet-Mongolian Paleontological Expeditions of the 1970s. It was named for the Garuda bird of Hindu mythology; its name means "short-footed Garuda mimic."**

145

GOYOCEPHALE

(GOY-oh-cee-FAL-ee)

The flat-headed *Goyocephale* was one of the most unusual pachycephalosaurs. It was found by the Joint Polish-Mongolian Paleontological Expeditions to the Gobi Desert and was named and described in 1982.

Few pachycephalosaurs are known from more than a skull, but some bones of the skeleton of *Goyocephale* were also found, including the tail and front and back limbs. The skeleton, although poorly preserved, shows that *Goyocephale* was built much like other pachycephalosaurs. The limbs were light, and the back was reinforced by bony tendons along the spines of the back bones. This strengthened the spinal cord and probably helped reduce stress during head-butting contests.

Goyocephale's prominent, caninelike teeth in the front of both the upper and lower jaws probably were a display to scare other animals away. This has also been found in some primitive deer (muntjacs) and the heterodontosaurid dinosaurs.

Goyocephale is one of the few known flat-headed pachycephalosaurs. Its relatives included *Homalocephale* and *Wannanosaurus,* both from the Late Cretaceous of central and eastern Asia.

Period:
Late Cretaceous

Order, Suborder, Family:
Ornithischia, Marginocephalia, Homalocephalidae

Location:
Asia (Mongolian People's Republic)

Length:
Probably no more than 6½ feet (2 meters)

A skeletal drawing of *Goyocephale.* One of the few known flat-headed pachycephalosaurs, it had lightly built limbs and a spine reinforced by bony tendons.

Goyocephale's prominent, caninelike teeth . . . probably were a display to scare other animals away.

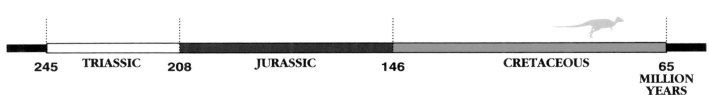

Goyocephale lattimorei. Although flat-headed, *Goyocephale* probably had head-butting contests in which males fought for females and territory.

| 245 | TRIASSIC | 208 | JURASSIC | 146 | CRETACEOUS | 65 MILLION YEARS AGO |

HADROSAURUS
(HAD-roh-SORE-us)

Joseph Leidy, a 19th-century scientist, was the first to realize that plant-eating dinosaurs such as *Hadrosaurus* did not resemble large, sprawling lizards. Instead, they stood more upright; Leidy thought they stood like kangaroos.

*H*adrosaurus, named and described in 1858, was the first dinosaur to have its skeleton mounted. Specimens of this dinosaur were displayed first at the Academy of Natural Sciences in Philadelphia, then at Princeton, the Smithsonian in Washington, D.C., and the Field-Columbia Museum in Chicago (now the Field Museum of Natural History).

Hadrosaurus stood on its back legs. Its front limbs were shorter than its legs. It probably had a horny beak for cropping plants and leaves, and behind its beak it had a complex tooth system for chewing. For many years, scientists had little idea what *Hadrosaurus* looked like. The skull was not found with the rest of the skeleton. Recent studies have shown that it looked like the hook-nosed hadrosaurids such as *Gryposaurus* and the less-well-known *Kritosaurus*. Both of these animals had a deep, narrow face with a rounded arch above the nostrils (it is not known if *Hadrosaurus* had this as well). This arched snout was probably covered with thick skin and may have been used both for display and as a fighting structure. If so, it may have been more prominent in males than in females.

These hook-nosed hadrosaurids were close relatives of *Hadrosaurus*. Because they are similar, they have often been confused. *Hadrosaurus* probably lived only in the eastern part of the United States. *Aralosaurus* from the former Soviet Union was also a close relative of these hook-nosed duckbilled dinosaurs.

Period:
Late Cretaceous

Order, Suborder, Family:
Ornithischia, Ornithopoda, Hadrosauridae

Location:
North America (United States)

Length:
30 feet (9 meters)

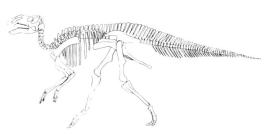

A skeletal drawing of *Hadrosaurus* ("thick reptile"), the first hadrosaur skeleton to be discovered. Most of it was found in marine rocks of Late Cretaceous age in southern New Jersey near Philadelphia.

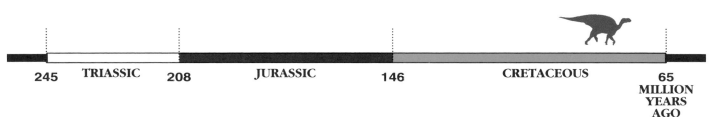

| 245 | TRIASSIC | 208 | JURASSIC | 146 | CRETACEOUS | 65 MILLION YEARS AGO |

HOMALOCEPHALE

(HOME-ah-loh-SEF-ah-lee)

As its name "even head" suggests, *Homalocephale* had a flat head unlike most pachycephalosaurs. The dinosaur is known from limited but very good material. The single skull of *Homalocephale* is missing the front of the snout, but is otherwise complete. This skull proves that the animal was a flat, thick-headed reptile. The back of the skull was high and hung slightly over the neck. The sides of the skull were slightly thickened and expanded. The top of the skull, as well as the cheek region and back of the skull, had spikes, bumps, and ridges of bone. The large eye sockets probably mean that *Homalocephale*, like many other pachycephalosaurs, had good vision.

Other information can be discovered from its skull. There was a large area for the nerve used for smelling. The size of this area suggests that *Homalocephale* had a good sense of smell. This is useful for any animal in the wild; it could have smelled a nearby predator and escaped before being seen.

The teeth of *Homalocephale* show wear; it did a lot of chewing. It probably was a browsing dinosaur, feeding mostly on leaves and probably not eating many fruits.

From the skeleton of *Homalocephale,* we know that the thorax, abdomen, and pelvic area were extremely broad, perhaps for a large gut that was needed to digest a diet of plants.

Period:
Late Cretaceous
Order, Suborder, Family:
Ornithischia, Marginocephalia, Homalocephalidae
Location:
Asia (Mongolian People's Republic)
Length:
10 feet (3 meters)

The large eye sockets probably mean that *Homalocephale*, like many other pachycephalosaurs, had good vision.

Top: Skeletal drawings of *Homalocephale.* *Bottom:* *Homalocephale calathocercos.* The spine and the flat head with its rear shelf suggest the animal engaged in head-butting contests over females and territory.

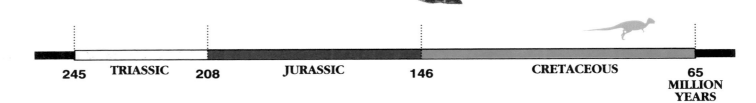

| 245 | **TRIASSIC** | 208 | **JURASSIC** | 146 | **CRETACEOUS** | 65 **MILLION YEARS AGO** |

HYPACROSAURUS

(hy-PAK-roh-SORE-us)

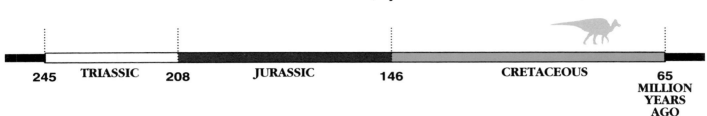

| 245 | TRIASSIC | 208 | JURASSIC | 146 | CRETACEOUS | 65 MILLION YEARS AGO |

The head of *Hypacrosaurus* looked much like *Corythosaurus*, another Late Cretaceous hadrosaurid. The snout was somewhat ducklike, although the nostrils were in slightly different places. The crest of *Hypacrosaurus* was shaped like a helmet, but more pointed at the top. The back of the skull was also similar.

The striking differences between the two dinosaurs were in the skeleton, especially the backbone. *Corythosaurus* had moderate-size spines on its back bones, but *Hypacrosaurus* had very tall ones. This is how the dinosaur got its name, which means "high-spined reptile." These tall spines may have allowed for extra back muscles, but they also formed a sail that would have captured the heat of the sun in the morning or released heat to cool the animal off on a hot day. In any case, these long spines would have made the animal look larger.

As in all other lambeosaurines, the crest of *Hypacrosaurus* contained a complex nasal cavity on top of its head. When it breathed, air was drawn in and out of the crest through a series of tubes and chambers that connected the nostrils to the back of the throat.

The animal could also make sounds with its crest. Because of the size and shape of the crest, the sounds produced would have been low notes. It probably used these sounds to communicate with other members of its herd.

A skeleton of *Hypacrosaurus*. A relative of *Corythosaurus*, it too had a crest containing a multipurpose nasal cavity, one that could have been used to produce sounds.

Period:
Late Cretaceous
Order, Suborder, Family:
Ornithischia, Ornithopoda, Hadrosauridae
Location:
North America (Canada, United States)
Length:
30 feet (9 meters)

Hypacrosaurus, or "high-spined reptile," was a duckbilled dinosaur whose sail not only made the animal look larger, but perhaps helped it regulate body temperature.

KRITOSAURUS

(KRIT-oh-SORE-us)

Kritosaurus was a large, flat-headed duckbilled dinosaur. It had a ridge of bone between the eyes and the snout that gave it a distinguished "Roman nose" appearance. In other ways, *Kritosaurus* was similar to its relative, *Hadrosaurus*. One of the larger hadrosaurs, *Kritosaurus* weighed up to three tons and stood around nine feet tall at the hips. The long, heavy tail gave this plant-eater balance when it walked on two legs. Its rear feet had three large toes ending in blunt hooves, like those of elephants. The front legs were small and not made for walking; the animal may have used them in gathering food or preparing nests.

The skull of *Kritosaurus* was long and slender, with a broad, flat face that looked like the bill of a duck. The ridge of bone in front of the eyes may have increased the nasal cavity to improve the sense of smell, an unusual feature for a reptile.

Kritosaurus had hundreds of tightly packed grinding teeth for crushing the tough plants it ate. To gather food, it probably stood on its tall rear legs and used its heavy tail as the third leg of a tripod. It then could reach high into trees to gather leaves and branches.

Like other duckbills, *Kritosaurus* probably lived in herds and protected its young from large predators such as *Albertosaurus*. *Kritosaurus* lived in southern North America from Texas to New Mexico. A close relative, *Gryposaurus*, lived in Alberta, Canada.

Like other duckbills, *Kritosaurus* probably . . . protected its young from large predators such as *Albertosaurus*.

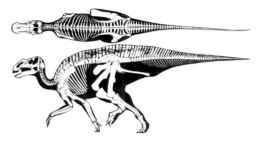

Skeletal drawings of *Kritosaurus*. The dinosaur was very likely bipedal, using its heavy tail for balance when it needed to stand on its rear legs.

Period:
Late Cretaceous
Order, Suborder, Family:
Ornithischia, Ornithopoda, Hadrosauridae
Location:
North America (United States)
Length:
30 feet (9 meters)

A Late Cretaceous forest scene, with three kritosaurs in the background. A plant-eater that moved in herds, *Kritosaurus* had a bony ridge on its snout that gave it a "Roman nose" appearance.

| 245 | TRIASSIC | 208 | JURASSIC | 146 | CRETACEOUS | 65 MILLION YEARS AGO |

LAMBEOSAURUS

(LAM-bee-oh-SORE-us)

245	TRIASSIC	208	JURASSIC	146	CRETACEOUS	65 MILLION YEARS AGO

A group of lambeosaurs. Closely related to *Corythosaurus*, *Lambeosaurus* lived concurrently and in the same places.

Period:	Late Cretaceous
Order, Suborder, Family:	Ornithischia, Ornithopoda, Hadrosauridae
Location:	North America (Canada, United States, Mexico)
Length:	30 feet (9 meters)

*L*ambeosaurus lived at the end of the Late Cretaceous. It was a hollow-crested hadrosaurid that lived at the same time and in the same places as *Corythosaurus* and *Parasaurolophus*.

One species of *Lambeosaurus* that was common in Alberta, Canada, had a crest that looked as if a hatchet were embedded in the top of its head. The front part of this crest was stout and pointed forward and up from just above the eyes. The crest was hollow and connected to the nostrils. The nasal cavity went down the back of the throat into the lungs. The hollow portion of the crest could produce low-frequency sounds for communication among family members or within large herds. The second part of the crest was farther back on its head and solid. This backward projecting "prong" probably supported a frill of skin that went along the back of the animal to the tail. It seems only males had this prong; only half the skulls found had them.

Because of the crests, paleontologists once thought *Lambeosaurus* and other hadrosaurids lived in water, using their crests as snorkels or for storing air. They may indeed have gone into the water or enjoyed a swim, but all hadrosaurids lived on land.

The name *Lambeosaurus*, coined in 1923, means "Lambe's reptile" in honor of its discoverer, Lawrence Lambe. Lambe described the first specimen of *Lambeosaurus* as *Stephanosaurus* in 1914; he did not realize it was a new genus. *Lambeosaurus* was closely related to *Corythosaurus* and *Hypacrosaurus*.

A skeleton of *Lambeosaurus lambei*. The dinosaur was robust, with long, sturdy front and back legs. The tail was long and was held straight back when the animal walked.

LEPTOCERATOPS

(LEP-toh-SAIR-ah-tops)

Leptoceratops, the first known protoceratopsid, was found along the Red Deer River of Alberta, Canada, in 1910. This skeleton with a partial skull was named *Leptoceratops gracilis*. It was the latest protoceratopsid, living to the end of the Cretaceous Period along with the large ceratopsid dinosaurs *Triceratops* and *Torosaurus*.

Leptoceratops was a lightly built protoceratopsid (its name means "slender-horned face") with long back limbs and short front limbs. The feet had tapered claws. *Leptoceratops* may have run on two legs or stood on its back legs to feed on tall vegetation. The large skull of *Leptoceratops* sloped down to a small, pointed, toothless beak. The dinosaur did not have nasal or brow horns. The neck frill was only slightly developed and solid, and it had a high, raised ridge down its midline.

More skeletons were discovered since the first was found. A partial one found in the St. Mary River Formation of Montana in 1916 was first thought to be a new species of *Leptoceratops;* however, it was later determined to be a new genus, so it was renamed *Montanaceratops cerorhynchus. Leptoceratops* and *Montanaceratops* are the only two known North American protoceratopsid dinosaurs.

Period:
Late Cretaceous
Order, Suborder, Family:
Ornithischia, Marginocephalia, Protoceratopsidae
Location:
North America (Canada, United States)
Length:
8 feet (2.4 meters)

One of only two known protoceratopsid dinosaurs from North America, *Leptoceratops* was lightly built, with a large skull and a poorly developed neck frill.

It was the latest protoceratopsid, living to the end of the Cretaceous Period along with the large ceratopsid dinosaurs *Triceratops* and *Torosaurus*.

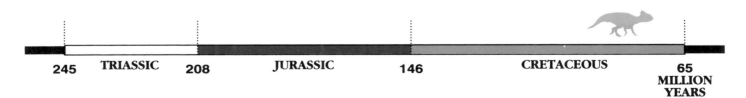

| 245 | TRIASSIC | 208 | JURASSIC | 146 | CRETACEOUS | 65 MILLION YEARS AGO |

OPISTHOCOELICAUDIA
(oh-PIS-thoh-SEE-lee-CAWD-ee-uh)

From the side, *Opisthocoelicaudia's* body was almost straight from the base of its neck to the beginning of its tail. . .

Period: Late Cretaceous
Order, Suborder, Family: Saurischia, Sauropodomorpha, Camarasauridae
Location: Asia (Mongolian People's Republic)
Length: 50 feet (15 meters)

Discovered by the Joint Polish-Mongolian Paleontological Expedition in 1965, the sauropod *Opisthocoelicaudia* is known from a skeleton with nearly all the bones of the body except the neck and head.

The name *Opisthocoelicaudia* refers to an unusual feature of the bones of the spine from the front half of the tail. The side of the spinal bone that faced the end of the tail curved deeply inward, and the side of the bone that faced the front of the animal curved deeply outward.

The dinosaur's name means "rear cavity tail." This may have allowed the animal to use its tail as a prop, like the third leg of a tripod, when it reached up for higher plants and trees. This would have helped balance the animal and taken weight off its hind legs. It also had an extra back bone in the pelvic area to strengthen the hips, and the hip socket was strong in order to hold the immense weight of this dinosaur when it stood on two legs.

The hips and tail were also unusual when the animal stood on its four feet. From the side, *Opisthocoelicaudia's* body was almost straight from the base of its neck to the beginning of its tail, unlike most sauropods. This shape was probably because of the specializations in the tail.

Most of the neck from the 1965 skeleton is missing, except for several bones. They show that, when the animal was moving, *Opisthocoelicaudia* held its neck with the head straight out in front of the body. The neck and skull of *Opisthocoelicaudia* may have been pulled apart by scavengers that left grooves on the pelvis and teeth marks on a bone in its leg.

A camarasaurid, *Opisthocoelicaudia* survived its Late Jurassic relatives in North America by 70 million years. It reached up to 50 feet in length and may have weighed as much as 20 tons.

245 **TRIASSIC** **208** **JURASSIC** **146** **CRETACEOUS** **65 MILLION YEARS AGO**

153

ORNITHOMIMUS

(or-NITH-oh-MIME-us)

This dinosaur has been found mainly in the Late Cretaceous Judith River and Horseshoe Canyon Formations of Alberta, Canada, but less complete specimens have been found in the western United States as well.

Ornithomimus ("bird mimic") looks much like a modern flightless bird, such as the ostrich. Both have small heads with beaks, long necks, and long back limbs for running fast. But *Ornithomimus* had a long tail and arms with three-fingered hands that ended in slender claws instead of short wings and no tail. The foot of ornithomimids was similar to the foot of some modern flightless birds; it had three toes on long upper foot bones. The ornithomimids may have been able to run up to 30 miles an hour.

Ornithomimus differs from its close relative *Struthiomimus* because its back and neck were shorter and its limbs were lighter.

All ornithomimids had lightly built skulls and no teeth. They may have been omnivorous, eating small vertebrates, insects, and fruits. But some paleontologists think they were only plant-eaters.

Ornithomimus ("bird mimic") looks much like a modern flightless bird, such as the ostrich.

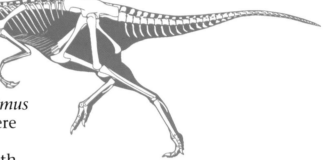

A skeletal drawing of *Ornithomimus*. The dinosaur had a long tail and limbs, three-fingered hands, and no teeth.

Period:
Late Cretaceous
Order, Suborder, Family:
Saurischia, Theropoda, Ornithomimidae
Location:
North America (Canada, United States)
Length:
13 feet (4 meters)

An *Ornithomimus altus* pair. The dinosaur is found mainly in two formations in Alberta, Canada. It may have eaten both plants and small animals.

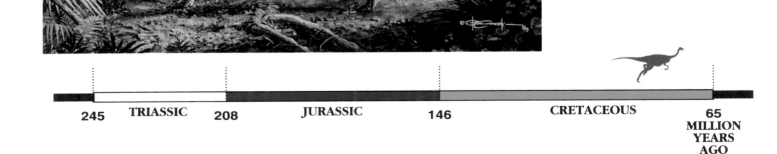

	TRIASSIC		JURASSIC		CRETACEOUS	
245		208		146		65 MILLION YEARS AGO

ORODROMEUS
(OR-oh-DROH-mee-us)

Period:
Late Cretaceous
Order, Suborder, Family:
Ornithischia, Ornithopoda, Hypsilophodontidae
Location:
North America (United States)
Length:
6½ feet (2 meters)

An *Orodromeus* nesting site. What makes *Orodromeus* special is that it is known from adult skulls and skeletons, younger animals, juveniles, hatchlings, and even an embryo. From all these fossils, we have a rich picture of the dinosaur's life history.

Orodromeus is a recently discovered dinosaur and one of the most spectacular. *Orodromeus* (the name means "mountain runner") was only about 6½ feet long as an adult.

The mother *Orodromeus* laid about 12 eggs in a tight spiral, with the first egg in the center of the spiral. These eggs were not large, only about 6 inches long and 2¾ inches wide. When the young were ready to come out of their eggs, they pecked through the top of the shell and climbed out of the nest. These hatchlings were nearly as fully developed as adults and could leave the nest and feed themselves. Possibly for protection, they stayed together after leaving the nest.

Hatchlings had long legs, with short front limbs and large heads. As they grew, their back limbs grew longer and their heads smaller in proportion to the rest of their body.

As an adult, *Orodromeus* was one of the longest-legged hypsilophodontids for its size. It must have been a very fast runner. It had a very long, straight tail, supported by many interwoven bony tendons. The tail balanced the front of the body when the dinosaur ran.

Orodromeus had simple, primitive teeth, much like the early ornithischian *Lesothosaurus*. Judging from the shape of these teeth, *Orodromeus* probably fed on fleshy fruits and possibly insects, especially when the animal was young.

Orodromeus comes from the Late Cretaceous Two Medicine Formation of western Montana. It was found with the hadrosaur *Maiasaura* and the meat-eating *Albertosaurus* and *Troodon*. *Orodromeus* may also have lived in southern Alberta, Canada.

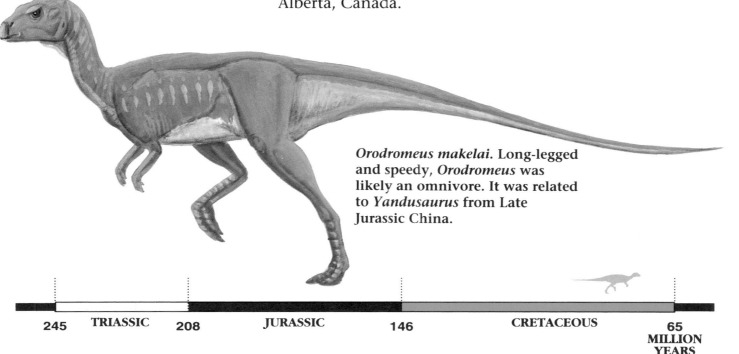

Orodromeus makelai. Long-legged and speedy, *Orodromeus* was likely an omnivore. It was related to *Yandusaurus* from Late Jurassic China.

| 245 | TRIASSIC | 208 | JURASSIC | 146 | CRETACEOUS | 65 MILLION YEARS AGO |

OVIRAPTOR
(OHV-ih-RAP-ter)

Period:
Late Cretaceous
Order, Suborder, Family:
Saurischia, Theropoda, Oviraptoridae
Location:
Asia (Mongolian People's Republic)
Length:
6 feet (1.8 meters)

The first specimen of *Oviraptor* was discovered by an American Museum of Natural History expedition to Asia in 1923. It was found lying next to a nest of eggs that may have belonged to *Protoceratops*. Perhaps the animal died while eating the eggs in the nest, killed by an angry *Protoceratops* parent.

This possible diet of eggs led to the genus and species name *Oviraptor philoceratops,* or "egg plunderer, lover of ceratopsians." At first, *Oviraptor* was thought to be a member of the Ornithomimidae, but its fingers were not the same length, and they ended in strongly curved claws. The ornithomimids had fingers about the same length and almost straight claws.

The skull of *Oviraptor* was short with large eye sockets, a crest above the snout, and a deep lower jaw with a large opening in the middle. Like the ornithomimids, both the upper and lower jaws were beaklike and toothless. The crest was full of holes or cavities, which were filled with air when it was alive.

Recently discovered specimens show that, aside from the skull, *Oviraptor* was similar to other theropods. An adult was about three feet tall at the shoulder and about six feet long from its nose to the tip of its tail. Its hands were moderately long, and the animal could use them to grasp. The long back legs show it was a good runner. With no teeth, *Oviraptor* may have had no other defense against other theropods than running.

Since the original specimen was discovered, several other skulls with skeletons have been found. These show that the size of the crest on each animal was different. The crest ranges from almost none to large. These differences are probably due to age.

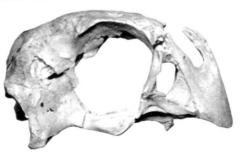

Above: Though *Oviraptor* was first found near a *Protoceratops* nest, it's likely, judging from its jaws, that the dinosaur was a plant-eater. *Left:* The upper half of an *Oviraptor* skull, showing the large eye socket and toothless beak.

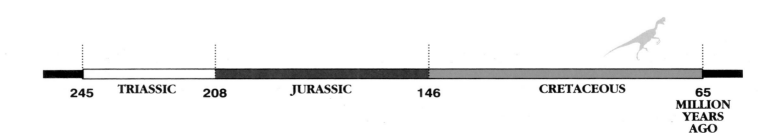

| 245 | TRIASSIC | 208 | JURASSIC | 146 | CRETACEOUS | 65 MILLION YEARS AGO |

PACHYCEPHALOSAURUS

(PACK-ee-CEF-ah-loh-SORE-us)

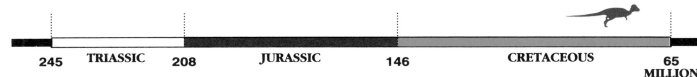

| 245 | TRIASSIC | 208 | JURASSIC | 146 | CRETACEOUS | 65 MILLION YEARS AGO |

The largest pachycephalosaur was *Pachycephalosaurus*. First found in rocks of Late Cretaceous age in Montana, *Pachycephalosaurus* was named and described by Barnum Brown and Eric Schlaikjer. *Pachycephalosaurus* is known only from a number of large domed skull roofs and a nearly complete skull. As the name *Pachycephalosaurus* ("thick-headed reptile") indicates, the top of the skull was very thick, almost nine inches in some skulls. Unlike other pachycephalosaurs, the snout stuck out, giving it an almost piglike profile. The top of the dome was smooth, much like that of *Prenocephale* and *Stegoceras*.

The back rim of the dome of *Pachycephalosaurus* and the top of the snout were covered with spikes and small horns. The teeth were simple, triangular blades. It ate soft plants.

Like their close relatives *Stegoceras, Prenocephale,* and *Stygimoloch, Pachycephalosaurus* males probably used their domed heads for head-butting contests to defend their territory and their females. The thick skull roof would have protected the small brain from damage.

The dome of a *Pachycephalosaurus* skull. In some cases, the bone was almost nine inches thick. The dome would have protected the brain during head-butting contests.

Period:	Late Cretaceous
Order, Suborder, Family:	Ornithischia, Marginocephalia, Pachycephalosauridae
Location:	North America (United States)
Length:	Estimated 15 feet (4.5 meters)

Pachycephalosaurus grangeri. A plant-eater, this dinosaur was first found in Montana in rock formations dating from the Late Cretaceous.

. . . Pachycephalosaurus **males probably used their domed heads for head-butting contests to defend their territory and their females.**

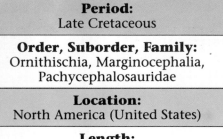

157

PACHYRHINOSAURUS
(PACK-ee-RHINE-oh-SORE-us)

Pachyrhinosaurus means "thick-nosed reptile"; the name refers to the bony facial pad, called the boss.

Pachyrhinosaurus was probably the most unusual and distinctive ceratopsid. It did not have brow or nasal horns; instead, it had a thick, bumpy, spongy pad of bone along the upper surface of its flattened face. This bony pad ran from the front of its nose back to above its eyes. The skull of *Pachyrhinosaurus* was massive; only *Triceratops, Pentaceratops,* and *Torosaurus* had larger skulls. *Pachyrhinosaurus* was the largest centrosaurine ceratopsian.

Charles Sternberg named *Pachyrhinosaurus canadensis* from three partial skulls found in southern Alberta, Canada. Because skulls of *Pachyrhinosaurus* are rare and because of their odd, gnarled facial pad or "boss," some paleontologists thought the facial pad was formed because a nasal horn broke off and then healed over. So they assumed the facial boss was a scar. Paleontologists recently uncovered a large bone bed with many *Pachyrhinosaurus* specimens in north central Alberta that proves that the facial pad is a normal feature and not a scar.

The rest of the skull of *Pachyrhinosaurus* looked very much like that of other centrosaurine ceratopsians such as *Centrosaurus, Monoclonius,* and *Styracosaurus.* Like them, *Pachyrhinosaurus* had a short frill, a deep face, and a short beak. *Pachyrhinosaurus* was also closely related to *Brachyceratops* and *Avaceratops.*

Period:
Late Cretaceous
Order, Suborder, Family:
Ornithischia, Marginocephalia, Ceratopsidae
Location:
North America (Canada)
Length:
20 feet (6 meters)

The skull of *Pachyrhinosaurus* was massive, and the dinosaur is considered the largest centrosaurine ceratopsian. It may also have been migratory: Pachyrhinosaur bones were recently found on Alaska's North Slope.

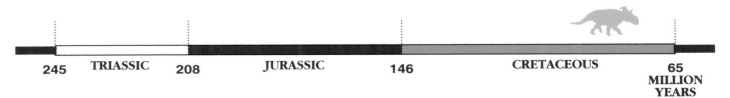

245	TRIASSIC	208	JURASSIC	146	CRETACEOUS	65 MILLION YEARS AGO

PANOPLOSAURUS
(pan-OP-loh-SORE-us)

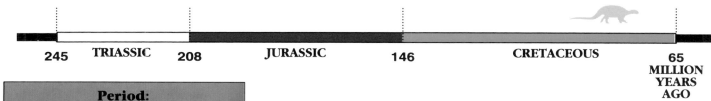

| 245 | **TRIASSIC** | 208 | **JURASSIC** | 146 | **CRETACEOUS** | 65 MILLION YEARS AGO |

Period:
Late Cretaceous

Order, Suborder, Family:
Ornithischia, Thyreophora, Nodosauridae

Location:
North America (Canada, United States)

Length:
23 feet (7 meters)

Panoplosaurus is known only from two partial skeletons, one of which preserves some of the armor the way it was in life. This skeleton shows that *Panoplosaurus* was unusual among nodosaurids because it did not have spikes on the sides of its neck.

It is not known how *Panoplosaurus* defended itself from a hungry *Daspletosaurus*. Perhaps the armor covering the rest of its body stopped the hungry predator from trying to attack, or perhaps *Panoplosaurus* used its tail to defend itself.

Panoplosaurus appears to be closely related to *Edmontonia*, and both lived at the same time. Both had large plates on their neck and smaller ridged plates on their body and tail. Their heads were similar, and they had large plates joined to the surface of their skulls. For further protection, the two dinosaurs had large oval plates covering their cheeks. It is not surprising that the dinosaur's name means "well-armored reptile."

The teeth of *Panoplosaurus* were small for a nodosaurid, and it had a diet of soft plants.

A *Panoplosaurus* skull. Known from only two partial skeletons, the dinosaur had relatively small teeth, which means it ate only soft plants.

Panoplosaurus was, as its name implies, a "well-armored reptile," with large and small plates over much of the body, including the cheeks, and a tail that may have been used defensively.

(The) skeleton shows that *Panoplosaurus* was unusual among nodosaurids because it did not have spikes on the sides of its neck.

159

PARASAUROLOPHUS
(PAIR-ah-SORE-ol-OH-fus)

Parasaurolophus walkeri. One of many North American hadrosaurs of the Late Cretaceous, the dinosaur was first thought to have been able to breathe under water. But that theory has been discredited, based on what scientists now know about the animal's crest.

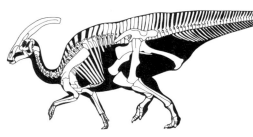

A skeletal drawing of *Parasaurolophus.* The trombonelike crest may have been used to make sounds—and even to help members of the species identify each other.

Parasaurolophus was an interesting-looking dinosaur. While the dinosaur seemed normal from the neck down, it looked almost as if it had a trombone on its head. And, in a way, it did.

Parasaurolophus is one of many hadrosaurs from the Late Cretaceous of North America. Its name, which means "like *Saurolophus,*" refers to the resemblance between the crests of these two duckbilled dinosaurs. However, the crest of *Saurolophus* was solid bone, while the crest of *Parasaurolophus* was hollow. The hollow space reached the nostrils and looped down to connect to the back of the throat. This crest was the animal's nasal cavity moved to the top of its head.

This crest, seen also in other lambeosaurine hadrosaurs, has attracted much attention. At first, paleontologists thought this crest was used under water, perhaps as a snorkel or a place to store air. Other suggestions have included extra space to increase the animal's sense of smell or an area used to cool its brain. The function of the crest is now thought to relate to hadrosaur social behavior. Because of the crest size and shape, it could have been for display. It may have helped other members of the species identify an animal, and it could have shown how old the animal was and its sex.

Also, because the crest was hollow and connected to the lungs, it would have made a resonating chamber. Sounds would have been made by a vocal organ or voice box and "pushed" through the crest, making a deep honking call. In this way, the animal could have communicated. All lambeosaurines would have used their "voices" to announce themselves, to warn their hatchlings, and to challenge other animals that invaded their territory.

Period:
Late Cretaceous
Order, Suborder, Family:
Ornithischia, Ornithopoda, Hadrosauridae
Location:
North America (Canada, United States)
Length:
33 feet (10 meters)

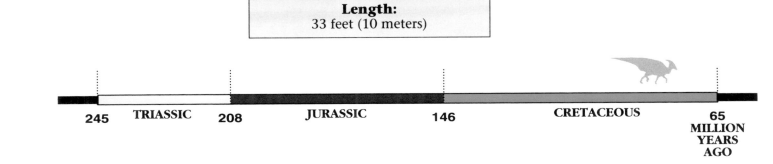

| 245 | TRIASSIC | 208 | JURASSIC | 146 | CRETACEOUS | 65 MILLION YEARS AGO |

PENTACERATOPS

(PEN-tah-SAIR-ah-tops)

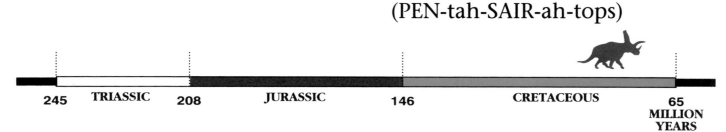

| 245 | TRIASSIC | 208 | JURASSIC | 146 | CRETACEOUS | 65 MILLION YEARS AGO |

Known only from the Late Cretaceous of northwestern New Mexico, *Pentaceratops* had one large horn on its snout, a pair of large horns above its eyes, and a pair of much smaller false horns in the cheek region. Its name means "five-horned face." The horns were actually bone.

Like other large ceratopsians, *Pentaceratops* resembled a rhinoceros in appearance and probably also in behavior and feeding habits. The skulls found in New Mexico were preserved with the different broad-leafed plants the animal ate. These plants resemble figs, willows, magnolias, and other types of hardwood flowering plants. *Pentaceratops* lived in thick forests where these plants grew.

This sturdy plant-eater stood eight feet tall at the shoulder and about ten feet tall from the ground to the top of the frill. Its skull was almost as long as that of *Torosaurus*. An adult *Pentaceratops* probably weighed as much as four or five tons. With the shield for protection and the horns to defend itself, this huge ceratopsian had few enemies.

| **Period:** |
| Late Cretaceous |
| **Order, Suborder, Family:** |
| Ornithischia, Marginocephalia, Ceratopsidae |
| **Location:** |
| North America (United States) |
| **Length:** |
| Estimated 25 feet (7.5 meters) |

Above: A skeletal drawing of *Pentaceratops. Left:* With its well-developed frill, "five-horned face," and sturdy body, *Pentaceratops* had little to fear from Late Cretaceous predators.

With the shield for protection and the horns to defend itself, this huge ceratopsian had few enemies.

PINACOSAURUS

(pie-NAK-oh-SORE-us)

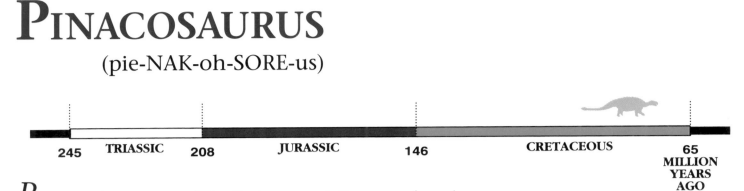

| 245 | TRIASSIC | 208 | JURASSIC | 146 | CRETACEOUS | 65 MILLION YEARS AGO |

Pinacosaurus was one of the first armored dinosaurs found in Asia. An expedition from the American Museum of Natural History went to Mongolia to search for traces of early humans; instead, it found dinosaur eggs and skeletons.

Scientists do not know what *Pinacosaurus* ate, but it and several other plant-eating dinosaurs were found in the same deposits, so some food was available. Recently, five juvenile skeletons of *Pinacosaurus* were found huddled together. They probably had been caught by a sandstorm and were buried alive.

The skull of *Pinacosaurus* is unusual because its entire surface was not covered with bone plates. Instead, there were few plates, covering only parts of the skull, and they were small.

But in all other aspects, *Pinacosaurus* looked like most other ankylosaurids. A collar of armor plates covered its neck, plates covered its body and tail, and it had a bony club on the end of its tail. Except for some unknown large teeth found with *Pinacosaurus*, its only predator was the small, sickle-clawed *Velociraptor*. But *Pinacosaurus* had little worry; a blow from its tail club could have killed its adversary.

Syrmosaurus was the name used by Russian paleontologists to describe *Pinacosaurus*.

Period:
Late Cretaceous
Order, Suborder, Family:
Ornithischia, Thyreophora, Ankylosauridae
Location:
Asia (People's Republic of China, Mongolian People's Republic)
Length:
16 feet (4.8 meters)

A *Pinacosaurus* skull. It had just a few small plates covering parts of the surface. Though scientists know the dinosaur was a herbivore, they do not know what kind of plants it ate.

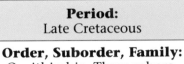

Pinacosaurus, called *Syrmosaurus* by Russian paleontologists, must have lived in a fairly dry or desert region because almost all of its fossils have been found in sand dune deposits.

PRENOCEPHALE
(PREN-oh-SEPH-a-lee)

An almost complete skull and part of a skeleton were found for *Prenocephale*. The fossils were well preserved and are some of the finest dinosaur material ever found. The fossil remains were collected during the Joint Polish-Mongolian Expeditions to the Gobi Desert. The animal was named and described in 1974.

The name *Prenocephale* means "sloping head," which refers to the large domed skull roof. The dome was high and rounded where it covered the braincase (the part of the skull that protects the brain). This dome was so large that it went back to the stout shelf, or frill, at the back of the skull. The animal had many rounded bumps and ridges of bone on the surface of its dome, face, and cheek. The bones of the skull were reinforced and tightly joined.

Prenocephale may have had very good eyesight; its eye sockets were large. Its teeth were simple; it was a plant-eater that probably fed on soft leaves or fruits. It also probably ate insects if it could catch them.

Like its relatives *Stegoceras* and *Pachycephalosaurus*, *Prenocephale* probably used its dome for head-butting contests. The thick skull roof would have protected the brain from damage during head-on collisions.

| **Period:** |
| Late Cretaceous |
| **Order, Suborder, Family:** |
| Ornithischia, Marginocephalia, Pachycephalosauridae |
| **Location:** |
| Asia (Mongolian People's Republic) |
| **Length:** |
| 6½ feet (2 meters) |

A *Prenocephale* skull. *Prenocephale* means "sloping head," referring to the large skull, rounded where it would have protected the brain. During head-butting contests, the extra padding would have come in handy.

Its teeth were simple; it was a plant-eater that probably fed on soft leaves or fruits. It also probably ate insects if it could catch them.

Prenocephale, with large eye sockets and probably very good eyesight, had rounded bumps and ridges of bone on its dome and face. The dinosaur was named and described in 1974.

245	**TRIASSIC**	208	**JURASSIC**	146	**CRETACEOUS**	65
						MILLION YEARS AGO

PROSAUROLOPHUS

(PROH-sore-OL-oh-fus)

Prosaurolophus was a common duckbilled dinosaur that lived during the Late Cretaceous in North America. It was discovered, named, and described by Barnum Brown of the American Museum of Natural History in 1916.

Prosaurolophus probably walked and ran on its stout, strong back legs. While resting, it supported itself with its more lightly built front limbs. The animal mostly used its front limbs for grasping and pulling branches, leaves, and fruit. Females may have used their front limbs for building nests.

The head of *Prosaurolophus* was much like other hadrosaurs. It had many teeth, with up to 50 teeth in each jaw. It also had as many as 250 replacement teeth in each jaw that were behind the top teeth and would replace the chewing teeth when they were needed.

The snout of *Prosaurolophus* looked much like a duck's. The rim of the beak would have been covered by a stiff bill for cropping foliage and other food. The most unusual part of its skull was the wide, solid bump above its eyes. The bump may have distinguished one animal from another, and males from females.

Prosaurolophus, which means "before *Saurolophus*," was closely related to the longer-crested *Saurolophus*. More distant relatives included *Edmontosaurus* and *Shantungosaurus*.

Above: A skeletal drawing of *Prosaurolophus*. Note the unusual bump just above the eye socket, which may have helped tell animals apart. *Below:* Two young albertosaurs chasing a *Prosaurolophus* and its brood.

Period:
Late Cretaceous
Order, Suborder, Family:
Ornithischia, Ornithopoda, Hadrosauridae
Location:
North America (Canada, United States)
Length:
30 feet (9 meters)

| 245 | TRIASSIC | 208 | JURASSIC | 146 | CRETACEOUS | 65 MILLION YEARS AGO |

SAICHANIA

(sah-ee-CHAIN-ee-uh)

245	TRIASSIC	208	JURASSIC	146	CRETACEOUS	65 MILLION YEARS AGO

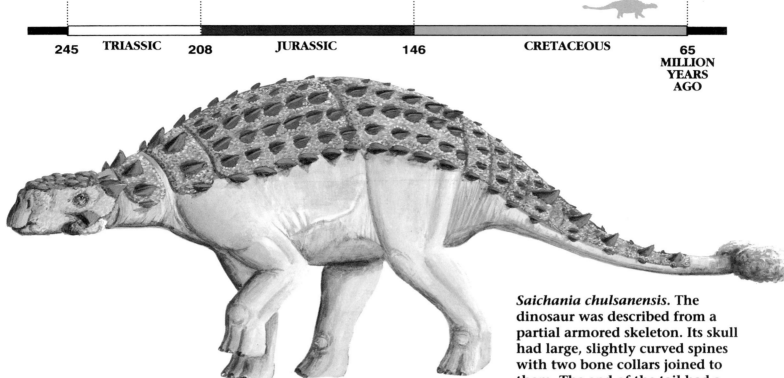

Saichania chulsanensis. The dinosaur was described from a partial armored skeleton. Its skull had large, slightly curved spines with two bone collars joined to them. The end of the tail had a small, plated club.

*S*aichania was described from a partial skeleton with the armor preserved the way it was when the animal was alive. The animal died in a sandstorm; it was found squatting on its belly in sandstone. At one time the entire skeleton was present, but erosion destroyed most of the rear. Several other partial skeletons recently have been discovered, and these add much to our knowledge of the animal.

The skull of *Saichania* had odd, domed armored plates joined to the top. The skull also had large spines that curved slightly down and back. It had two bone collars with ridged plates fused to them. Similar plates probably covered the body and tail. The end of the tail had a small club with three plates.

The teeth were small and suited to a diet of soft plants. Scientists do not know what these plants were, since the sandstone that preserved *Saichania* was poor at saving plants. These plants must have been suited to a hot, dry environment. To cope with this, *Saichania* may have moistened the air it breathed with a complex air passage in the skull. If the air was not moistened, then the lungs and other tissue would have lost water, causing the animal's death.

Living in such a hot environment also required *Saichania* to get rid of excess body heat. This would have been lost by evaporation of moisture in its air passage, much like a modern dog that pants on a hot day.

Period:
Late Cretaceous
Order, Suborder, Family:
Ornithischia, Thyreophora, Ankylosauridae
Location:
Asia (Mongolian People's Republic)
Length:
22 feet (6.6 meters)

. . . Saichania **may have moistened the air it breathed with a complex air passage in the skull. If the air was not moistened, then the lungs and other tissue would have lost water, causing the animal's death.**

165

SALTASAURUS

(SALT-ah-SORE-us)

From Salta in Argentina, *Saltasaurus* was described from several incomplete skeletons, none of which were found with a skull. This medium-size sauropod was dwarfed by its giant relative *Antarctosaurus,* also from South America.

The armor of *Saltasaurus* consisted of hundreds of small bones about the size of peas, tightly packed in the skin, and a few large, oval bony plates. The larger plates were about as broad and thick as a human palm. This bony protection seems to have covered the back and sides of its body and probably gave it a roughened, bumpy appearance.

Sauropod dinosaurs were not common in most parts of the world by the end of the Cretaceous Period. This may be because they were competing with more advanced plant-eating dinosaurs and because predatory dinosaurs kept their numbers down. Some predators had become very large (*Tyrannosaurus* weighed up to seven tons), and the size of sauropods was not as much protection as when predators were smaller. The armor on the titanosaurids may be what allowed *Saltasaurus* and other sauropods in this family to survive.

Saltasaurus and its relatives were successful in South America, and one member of the family established a population in North America. *Alamosaurus* in New Mexico and Utah faced the giant predators *Albertosaurus* and *Tyrannosaurus.*

The armor of *Saltasaurus* consisted of hundreds of small bones about the size of peas, tightly packed in the skin, and a few large, oval bony plates.

Period:
Late Cretaceous
Order, Suborder, Family:
Saurischia, Sauropodomorpha, Titanosauridae
Location:
South America (Argentina)
Length:
40 feet (12 meters)

A large predator attacking a 20-ton sauropod would have driven its claws deep into the flesh near the hips. The sudden blow would have harmed most sauropods, but not *Saltasaurus*. An armored sauropod was rare among dinosaurs—sauropods had lived for more than 70 million years, with size as their only defense.

| 245 | TRIASSIC | 208 | JURASSIC | 146 | CRETACEOUS | 65 MILLION YEARS AGO |

SAUROLOPHUS
(sore-OL-oh-FUS)

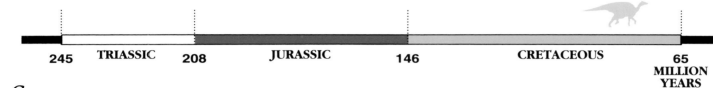

245	TRIASSIC	208	JURASSIC	146	CRETACEOUS	65 MILLION YEARS AGO

Saurolophus ("ridged reptile") was a hadrosaurid. It had a large, bony spike pointing back over the top of its head between its eyes. The front of this thin crest covered a shallow hole that went down to the nostril area of the snout. This hole may have been covered with a long, fleshy "bag" of skin, which the animal may have inflated to make loud, honking sounds. The bag may have also been a colorful display. Both features would have been useful during breeding season. The dinosaur could call its mate and let another male *Saurolophus* know that it was the dominant male of the herd.

The skeleton of *Saurolophus* was much like that of other hadrosaurids. The back limbs were long and well built, while the front limbs were shorter. The tail was long and held high off the ground, acting as a balance when *Saurolophus* walked or ran on its back legs.

This Late Cretaceous hadrosaurid has been found in southern Alberta, Canada, and in Mongolia. There may have been a land connection that allowed this animal to live in central Asia and North America.

Saurolophus was closely related to *Prosaurolophus* and less closely to *Edmontosaurus*.

A *Saurolophus* skull. The duckbilled dinosaur, a cousin of *Edmontosaurus*, had a shallow hole in its snout that may have played a role in making honking sounds.

Period:
Late Cretaceous

Order, Suborder, Family:
Ornithischia, Ornithopoda, Hadrosauridae

Location:
North America (Canada), Asia (Mongolian People's Republic)

Length:
33–43 feet (10–13 meters)

Saurolophus, whose fossil remains have been found in both Canada and Mongolia, had long back limbs, shorter front ones, and a long tail held off the ground. Scientists speculate that a land bridge between Asia and North America might have allowed *Saurolophus* to migrate between the two continents.

167

SAURORNITHOIDES
(SORE-orn-ith-OID-eez)

Discovered during an expedition by the American Museum of Natural History, *Saurornithoides* is an example of a birdlike dinosaur. It was found close to where two other birdlike dinosaurs, *Velociraptor* and *Oviraptor,* were also found.

This agile, 30-pound predator had a long, slender snout and large eyes set deep in their sockets. The front limbs of *Saurornithoides* were long for a dinosaur predator that walked on two legs. On each hand were three fingers that ended in sharp claws. The back limbs had feet with four toes each. One toe pointed down, two large toes pointed forward, and the inner toe was held off the ground. This inner toe had a sickle-shaped claw that may have been used as a weapon. A relative, *Deinonychus,* also had a sickle-shaped claw on its foot.

The teeth of *Saurornithoides* were different from its closest relative, *Troodon.* Usually, the teeth of animals in the family Troodontidae had large, round roots and jagged edges like those found on steak knives. The jagged edges on the teeth of *Saurornithoides* were only on one side, and there were no such edges at the tips.

Saurornithoides looked somewhat like a modern flightless bird. Its name means "birdlike reptile." It may have even had a similar lifestyle. It searched for prey, perhaps during twilight or even at night. *Saurornithoides* may have preferred small mammals, although it probably ate any prey it could catch.

Period:
Late Cretaceous
Order, Suborder, Family:
Saurischia, Theropoda, Troodontidae
Location:
Asia (Mongolian People's Republic)
Length:
6½ feet (2 meters)

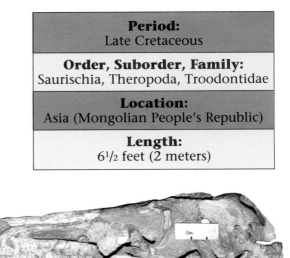

A *Saurornithoides* skull. This birdlike predator, with its long, slender snout and large, deep-socketed eyes, may have hunted at twilight, eating anything it could catch.

Saurornithoides looked somewhat like a modern flightless bird. Its name means "birdlike reptile."

Saurornithoides mongoliensis. This 30-pound predator had back feet with four toes each. The inner toe was held off the ground and had a sickle-shaped claw similar to the claw on *Deinonychus's* foot. *Saurornithoides's* teeth were partially jagged, or serrated, like a steak knife.

245	TRIASSIC	208	JURASSIC	146	CRETACEOUS	65 MILLION YEARS AGO

SEGNOSAURUS
(SEG-noh-SORE-us)

Segnosaurus galbinensis, or "slow lizard from Galbin" (a region of the Gobi Desert), was first described by Mongolian paleontologist Altangerel Perle in 1979. It was an unusual saurischian that Perle classified in its own family, the Segnosauridae.

The dinosaur had an unusual combination of features. Its pelvis looked much like the pelvis of the dromaeosaurids, although it was much larger. *Segnosaurus* had feet with long, slender, theropodlike claws and ankles, although it had four toes instead of three on each foot. The teeth, although numerous and small, resembled those of some theropods.

The pelvis of *Segnosaurus* was very wide, giving the animal a broad back and a "pot belly." By contrast, the pelvis of most theropods was slender. *Segnosaurus's* feet were not really theropod feet. Its relative *Erlikosaurus* had a prosauropodlike beak, so *Segnosaurus* probably did, too.

The lifestyle of *Segnosaurus* and its relatives is not known. The teeth show they were probably plant-eaters. They may have had a large gut for digesting plants and looked like large, wide-bodied prosauropods, probably walking on all four legs most of the time. They may have been bearlike or perhaps resembled extinct Ice Age mammals called ground sloths.

Segnosaurians were probably not fast runners. If their hands did have claws, they would have been used as weapons against predators, including *Alectrosaurus.*

The pelvis of *Segnosaurus* was very wide, giving the animal a broad back and a "pot belly."

Period:	Late Cretaceous
Order, Suborder, Family:	Saurischia, Segnosauria, Segnosauridae
Location:	Asia (Mongolian People's Republic)
Length:	Estimated 20 feet (6 meters)

Segnosaurus galbinensis. One of the most important finds of the Joint Soviet-Mongolian Paleontological Expeditions of the 1970s was a new group of dinosaurs, the segnosaurians. Segnosaurs were probably plant-eaters, with a large gut for digesting vegetation. They may have walked on all fours, resembling bears.

245	TRIASSIC	208	JURASSIC	146	CRETACEOUS	65 MILLION YEARS AGO

SHANSHANOSAURUS

(shan-SHAN-oh-SORE-us)

The paleontological expeditions into China's Turpan Basin in 1964–66 turned up several interesting and unusual dinosaurs. In the Subashi Formation, workers discovered the incomplete skeleton of a new, small theropod. It was described in 1977 and named *Shanshanosaurus huoyanshanensis*. An adult animal, it proved to be so different from other Chinese theropods that a new family, Shanshanosauridae, was created for it.

The footlong skull of *Shanshanosaurus* had large eye sockets and even larger openings in front of the eye sockets, which made the skull light. In many ways, the skeleton suggests the dinosaur may have been related to the large theropods, such as the tyrannosaurids, but other features suggest that it was related to the dromaeosaurids. Some scientists have noted that the front teeth of *Shanshanosaurus* look like the teeth of *Aublysodon* and have suggested the two dinosaurs may be related.

Shanshanosaurus hunted small animals, such as lizards and small mammals. The large eye sockets may mean it was active after the sun went down.

In many ways, the skeleton suggests the dinosaur may have been related to the large theropods, such as the tyrannosaurids, but other features suggest that it was related to the dromaeosaurids.

Shanshanosaurus huoyanshanensis. The dinosaur, discovered in the mid-1960s in China's Subashi Formation, had large eye sockets. Perhaps *Shanshanosaurus* was nocturnal, preying on small reptiles and mammals.

Period:
Late Cretaceous
Order, Suborder, Family:
Saurischia, Theropoda, Unknown
Location:
Asia (People's Republic of China)
Length:
6–10 feet (1.8–3 meters)

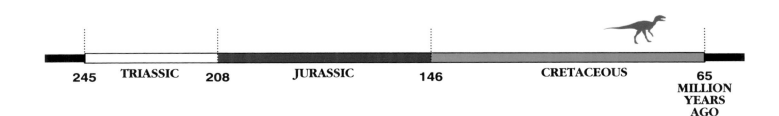

| 245 | TRIASSIC | 208 | JURASSIC | 146 | CRETACEOUS | 65 MILLION YEARS AGO |

SHANTUNGOSAURUS
(shan-TUNG-oh-SORE-us)

Shantungosaurus giganteus.

A skeletal drawing of *Shantungosaurus.* The quadruped "reptile from Shandong," known from several unjoined bones, had a long skull with jaws that probably held the most teeth of any dinosaur—about 1,400!

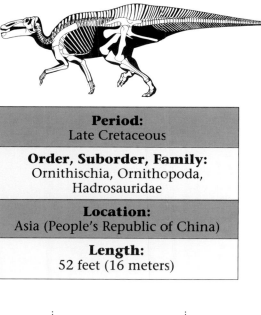

Period:
Late Cretaceous
Order, Suborder, Family:
Ornithischia, Ornithopoda, Hadrosauridae
Location:
Asia (People's Republic of China)
Length:
52 feet (16 meters)

Shantungosaurus may have been the largest hadrosaur. It is larger than some of the smaller sauropods. Named and described in 1973, *Shantungosaurus* is known from many unjoined bones from the Shandong Province, People's Republic of China.

Shantungosaurus was a flat-headed hadrosaur, much like its North American relative, *Edmontosaurus.* The skull was long, with an extended snout and jaws. The jaws had a lot of room for many teeth—some 1,400 teeth in all. This was probably the most seen in any dinosaur. The muscles for the jaw connected to a large area at the back of the skull.

Around the nostrils was a large hole where there was probably a fleshy "bag." The animal may have inflated the bag to make sounds or to attract a mate. It may have also used it to defend its territory.

The "reptile from Shandong" had very stout back limb bones, which it needed to support its large size. *Shantungosaurus* probably used both its front and back legs to walk, even when trying to escape from its enemy, *Tarbosaurus.*

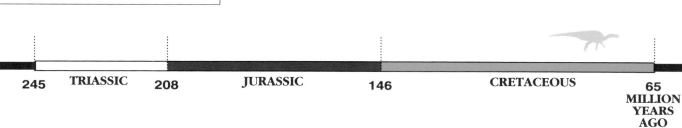

| 245 | TRIASSIC | 208 | JURASSIC | 146 | CRETACEOUS | 65 MILLION YEARS AGO |

STEGOCERAS
(ste-GOSS-er-as)

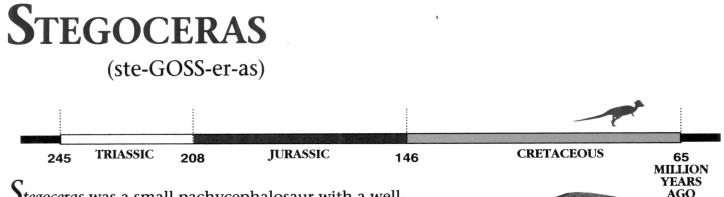

| 245 | TRIASSIC | 208 | JURASSIC | 146 | CRETACEOUS | 65 MILLION YEARS AGO |

Stegoceras was a small pachycephalosaur with a well-developed domed skull roof. It was closely related to the domed pachycephalosaurs, including *Pachycephalosaurus, Prenocephale,* and *Stygimoloch,* and was the first member of the family Pachycephalosauridae to be found.

We know much about *Stegoceras* because of a good skeleton (the only other pachycephalosaur preserved with its skeleton is *Homalocephale*) and many partial skulls of young and adult individuals.

The dome of *Stegoceras* was flat in young animals but became large and thick in adults. Since there are many *Stegoceras* domes that have become part of the fossil record, we know there are two kinds of domes among adults. The thicker, heavier domes may have been male skulls, while the thinner, lower domes perhaps belonged to females.

Stegoceras may have used its thick dome for head-butting contests. Males would have held these contests to win females or territory. This would also explain why the male skull domes were thicker.

There is other evidence that these animals had head-butting contests. The braincase, back of the skull, and backbone all show that forces were sent from the dome through the head, around the braincase, and down the backbone to the limbs. In this way, animals like *Stegoceras* could survive the stress of head-to-head combat, much the same way as goats and sheep do today.

A *Stegoceras* skull. Closely related to *Pachycephalosaurus,* the dinosaur had a domed skull that was flat in juveniles, but large and thick in adults. The thickness was probably useful in head-butting contests.

Period:
Late Cretaceous
Order, Suborder, Family:
Ornithischia, Marginocephalia, Pachycephalosauridae
Location:
North America (Canada, United States)
Length:
8 feet (2.5 meters)

Stegoceras validus. The first pachycephalosaur discovered, it was described in 1902. *Stegoceras* was thought to be a ceratopsian; however, later material suggested it was related to the stegosaurs. Then a skull and partial skeleton were described in 1924 by Charles Gilmore, who said it was *Troodon.* With the discovery of *Pachycephalosaurus* in 1943, *Stegoceras* was correctly classified.

STRUTHIOMIMUS
(STRUTH-ee-oh-MIME-us)

Struthiomimus ("ostrich mimic") is the best known of all the ornithomimids. A complete skeleton, which is now displayed at the American Museum of Natural History, was collected from the Judith River Formation of Alberta, Canada. Its name suggests a resemblance to the modern *struthio* (ostrich).

Like other ornithomimids, the skull was small (only ten inches long) and lightly built; it lacked teeth and had a horny beak. The eyes were large and the neck was slender. Just as in modern birds, the neck ribs were solidly joined to the neck bones. The back was stiff to support the weight of the body, long arms, and neck.

The arms were slender and the hands had three fingers, the inner one slightly shorter than the others. The claws at the ends of the fingers were straight and were probably not used to grasp prey. The back limbs and upper foot bones were long, as was the tail. The hands and arms of *Struthiomimus* were like those of a modern sloth, which uses its arms to grasp branches of trees.

Because of this, some scientists think *Struthiomimus* may have been a plant-eater that used its arms to pull fruit off branches within reach of its beak. But it may have been omnivorous and eaten plants and whatever small animals it could catch.

Period:
Late Cretaceous
Order, Suborder, Family:
Saurischia, Theropoda, Ornithomimidae
Location:
North America
Length:
13 feet (4 meters)

The hands and arms of *Struthiomimus* were like those of a modern sloth, which uses its arms to grasp branches of trees.

Struthiomimus, **from Late Cretaceous Canada, is somewhat similar to the modern ostrich. Scientists are unsure whether it was a herbivore or whether it ate both plants and small animals.**

| 245 | TRIASSIC | 208 | JURASSIC | 146 | CRETACEOUS | 65 MILLION YEARS AGO |

STYRACOSAURUS

(sty-RACK-oh-SORE-us)

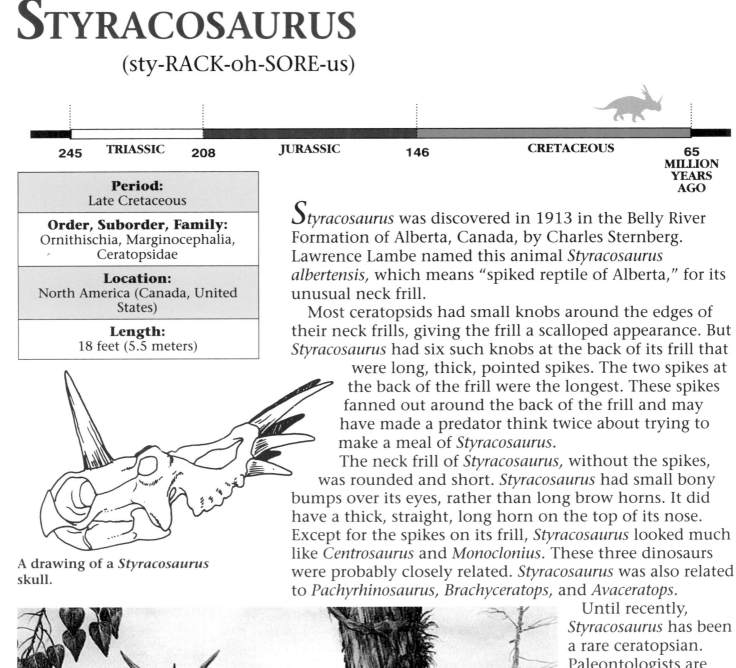

245	TRIASSIC	208	JURASSIC	146	CRETACEOUS	65 MILLION YEARS AGO

Period:
Late Cretaceous
Order, Suborder, Family:
Ornithischia, Marginocephalia, Ceratopsidae
Location:
North America (Canada, United States)
Length:
18 feet (5.5 meters)

Styracosaurus was discovered in 1913 in the Belly River Formation of Alberta, Canada, by Charles Sternberg. Lawrence Lambe named this animal *Styracosaurus albertensis*, which means "spiked reptile of Alberta," for its unusual neck frill.

Most ceratopsids had small knobs around the edges of their neck frills, giving the frill a scalloped appearance. But *Styracosaurus* had six such knobs at the back of its frill that were long, thick, pointed spikes. The two spikes at the back of the frill were the longest. These spikes fanned out around the back of the frill and may have made a predator think twice about trying to make a meal of *Styracosaurus*.

The neck frill of *Styracosaurus*, without the spikes, was rounded and short. *Styracosaurus* had small bony bumps over its eyes, rather than long brow horns. It did have a thick, straight, long horn on the top of its nose. Except for the spikes on its frill, *Styracosaurus* looked much like *Centrosaurus* and *Monoclonius*. These three dinosaurs were probably closely related. *Styracosaurus* was also related to *Pachyrhinosaurus, Brachyceratops,* and *Avaceratops.*

A drawing of a *Styracosaurus* skull.

Until recently, *Styracosaurus* has been a rare ceratopsian. Paleontologists are now studying a number of *Styracosaurus* skeletons that have been found in Montana. These new skeletons may prove to be a new species.

Two styracosaurs. Closely related to *Centrosaurus,* *Styracosaurus* is known for its heavily spiked neck frill. Once thought to be a rare ceratopsian, *Styracosaurus* now is known to have lived in Late Cretaceous Montana as well as Canada.

TARCHIA
(TAR-kee-uh)

Tarchia giganteus is known from a complete skull and a partial skeleton; however, except for the skull, little of the material has been described. For this reason, neither the length nor the weight of the animal is known.

The skull of *Tarchia* was massive, and it had room for a large brain. In fact, its name means "brain" in Mongolian. The skull had large spinelike plates at the top and bottom. Large, knobby plates of armor covered the skull. The beak was broad and rounded. The teeth looked like those of most other ankylosaurs; the animal was a plant-eater. The armor was thin-walled and similar to the armor plates of most ankylosaurs.

Tarchia lived in a dry environment. Other dinosaurs that lived at the same time and in the same place included *Tarbosaurus*, the ostrich dinosaur *Gallimimus*, and the hadrosaur *Saurolophus*.

Scientists do not know what plants *Tarchia* ate because no fossil plants have been found.

Period:
Late Cretaceous
Order, Suborder, Family:
Ornithischia, Thyreophora, Ankylosauridae
Location:
Asia (Mongolian People's Republic)
Length:
Unknown

The dinosaur lived in a dry climate, and its teeth looked like most ankylosaurs'. *Tarchia* was a plant-eater, though scientists do not know what vegetation the dinosaur may have eaten since no fossil plants have survived with the dinosaur's remains.

The skull of *Tarchia* was massive, and it had room for a large brain. In fact, its name means "brain" in Mongolian.

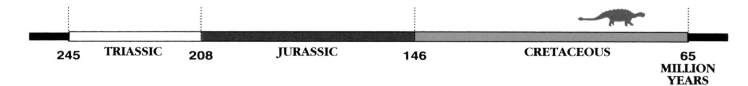

| 245 | **TRIASSIC** | 208 | **JURASSIC** | 146 | **CRETACEOUS** | 65 **MILLION YEARS AGO** |

THESCELOSAURUS

(THESS-ah-loh-SORE-us)

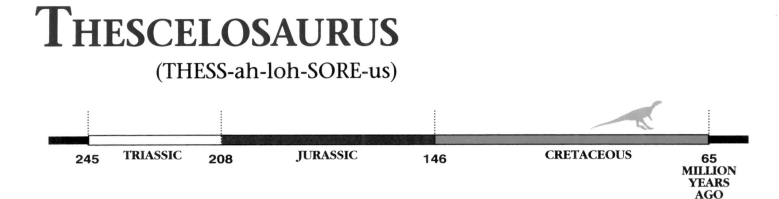

245 TRIASSIC **208**	JURASSIC	**146**	CRETACEOUS **65** MILLION YEARS AGO

A skeleton of *Thescelosaurus neglectus*. This "beautiful reptile" is known from several partial skeletons and skulls found in Late Cretaceous rocks in Montana and two Canadian provinces.

Thescelosaurus was one of the last hypsilophodontid dinosaurs. It is known from the end of the Late Cretaceous of Montana in the United States and the provinces of Alberta and Saskatchewan in Canada. *Thescelosaurus* may have seen what happened to cause the mass extinctions at the end of the Mesozoic Era.

Meaning "beautiful reptile," *Thescelosaurus* is known from several partial skeletons and incomplete skulls. The front of the jaws lacked teeth. Muscular cheeks along the sides of the face kept food from falling out of the mouth during chewing.

This small dinosaur had stocky body and limb proportions. *Thescelosaurus* must have been a much slower runner than other hypsilophodontids that were about the same size.

Thescelosaurus may have seen what happened to cause the mass extinctions at the end of the Mesozoic Era.

Period: Late Cretaceous
Order, Suborder, Family: Ornithischia, Ornithopoda, Hypsilophodontidae
Location: North America (Canada, United States)
Length: 11 feet (3.3 meters)

Small and stocky, *Thescelosaurus* had muscular cheeks for holding food while it chewed. One of the last hypsilophodontids, *Thescelosaurus* may not have been as swift a runner as similar-size members of the family.

TOROSAURUS
(TORE-oh-SORE-us)

The first two *Torosaurus* specimens were a pair of skulls found in Wyoming in 1891 by John Bell Hatcher. They were described by Othniel Marsh later that same year. These two skulls were named *Torosaurus latus* and *Torosaurus gladius.*

Torosaurus's skull had a large, flaring neck frill that was more than six and a half feet long and about as wide; it had the largest skull of any known land animal. (A recently discovered skull measures nearly nine feet in total length.) The size of the skull, along with the two very long and robust brow horns, were the inspiration for the name *Torosaurus,* which means "bull reptile."

Although very large, the neck frill of *Torosaurus* was thin, and the two openings in the frill were large and oval-shaped. Some specimens show extra openings along the sides of the neck frill. *Torosaurus* had a small, drawn-out beak topped by a short nasal horn. Very little is known about the rest of the skeleton.

Torosaurus lived at the same time as its close relative *Triceratops.* But while *Triceratops* is common, *Torosaurus* is rare and may not have existed in such great numbers as *Triceratops.*

Torosaurus was also closely related to *Anchiceratops, Arrhinoceratops, Pentaceratops,* and *Chasmosaurus.*

Period:
Late Cretaceous
Order, Suborder, Family:
Ornithischia, Marginocephalia, Ceratopsidae
Location:
North America (Canada, United States)
Length:
25 feet (7.5 meters)

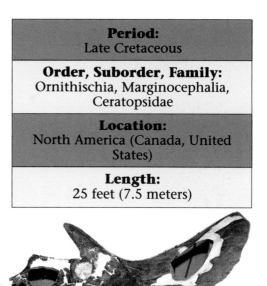

A *Torosaurus* skull. Though very large, the flaring neck frill was rather thin. Two torosaur skulls found in 1891 and named as different species are now thought to be a single species, *Torosaurus latus.*

Torosaurus, which lived at the same time as *Triceratops* but in smaller numbers, had a large skull and two long brow horns that gave the dinosaur its name, "bull reptile."

245	TRIASSIC	208	JURASSIC	146	CRETACEOUS	65 MILLION YEARS AGO

TROODON
(TROH-oh-don)

Period:
Late Cretaceous
Order, Suborder, Family:
Saurischia, Theropoda, Troodontidae
Location:
North America (Canada, United States)
Length:
8 feet (2.4 meters)

Troodon chasing its prey. With a well-developed brain and more teeth than any other theropod, *Troodon* must have proved a fierce enemy.

*T*roodon was described in 1856 by Joseph Leidy on the basis of a single small tooth. It was one of the first North American dinosaurs studied.

Many scientists believed the original *Troodon* tooth belonged to a theropod. *Troodon* was placed in the family Troodontidae in 1948. Then, in 1983, Jack Horner found the lower jaw of a small theropod with the same type of teeth, proving that *Troodon* was a theropod. It was also found that all the teeth of *Troodon* were not the same; the shape of the tooth depended on where the tooth was in the jaw. This jaw also showed that the animal called *Stenonychosaurus* and another named *Pectinodon* were the same as *Troodon*. A close relative of *Troodon* was *Saurornithoides*.

Troodon is now known from several partial skulls and skeletons. These skulls show that it had a large brain for its size, and it probably had the most developed brain of any dinosaur. Its eyes were large, taking up a big part of the skull. Each side of the lower jaw had 35 teeth—more than any other theropod.

Troodon's hands had slender fingers, and the inner finger ended in a large, thin, sharply pointed claw. The foot was similar to the foot of dromaeosaurids, with a large claw on the second toe. The claw was used to slash its prey. However, it was higher on the foot and smaller than the claw of a dromaeosaur. The claw on the first finger of the hand of *Troodon* was much larger than the largest claw on its foot. Also, like the dromaeosaurids, the back part of the tail was stiff. The dromaeosaurs may have been the ancestors of the troodontids.

Troodon formosus. With features like dromaeosaurids, *Troodon* had hands with slender fingers, and the inner finger ended in a sharp claw. The foot had a large claw on the second toe that was used to slash prey.

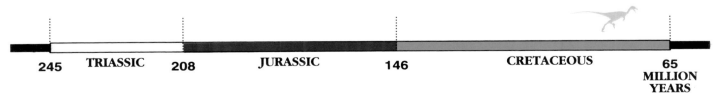

| 245 | **TRIASSIC** | 208 | **JURASSIC** | 146 | **CRETACEOUS** | 65 **MILLION YEARS AGO** |

VELOCIRAPTOR
(vel-OS-ih-RAP-ter)

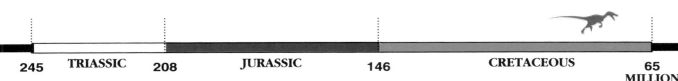

| 245 | TRIASSIC | 208 | JURASSIC | 146 | CRETACEOUS | 65 MILLION YEARS AGO |

The most amazing find in Mongolia may be the discovery of the skeletons of the small theropod *Velociraptor* ("speedy predator") with its right arm clamped firmly in the beak of *Protoceratops*. Both skeletons are complete. They are a picture of a Late Cretaceous struggle to the death. Soon after their deaths, they were buried by the drifting sands of a dune. They lay together in this death pose until 1971, when they were unearthed.

In 1923, the first specimen of *Velociraptor* was found by the American Museum of Natural History. Like the famous death-pose specimen, it was found in Late Cretaceous sandstone of the Djadokhta Formation in the Gobi Desert. And it was also found lying alongside a skull of *Protoceratops*.

Velociraptor was a small theropod, with a large, sickle-shaped claw on the second toe of its foot. It had a low, narrow snout, which is different from other members of its family. The jaws were lined with jagged teeth for tearing flesh. It swallowed its food in gulps instead of chewing, like most theropods. The arms were long, and it had strong chest and arm muscles.

Since the death-pose specimen was found with *Protoceratops*, *Velociraptor* probably ate this small ceratopsian, but it may have hunted larger prey. Its diet also included small animals, such as lizards.

Another dromaeosaurid feature that can be clearly seen in the death-pose specimen is the long pieces of bone that stiffen the tail, letting it act as a balance when the animal walked or ran. The tail, however, was still flexible, especially where it was attached to the hips.

Since the death-pose specimen was found with *Protoceratops*, *Velociraptor* probably ate this small ceratopsian, but it may have hunted larger prey.

Period:
Late Cretaceous

Order, Suborder, Family:
Saurischia, Theropoda, Dromaeosauridae

Location:
Asia (Mongolian People's Republic)

Length:
6 feet (1.8 meters)

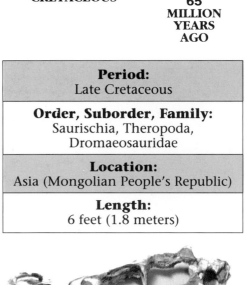

Above: A *Velociraptor* skull. *Below:* *Protoceratops* defending its nest against *Velociraptor*. A dangerous predator, *Velociraptor* was strong, long-limbed, and equipped with jagged cutting teeth. Though small, it was a "speedy predator."

CHAPTER 7

DINOSAUR EXTINCTION

Dinosaurs became extinct at the end of the Cretaceous Period. How and why are a puzzle that paleontologists try to solve by studying fossils and rock formations. These do not supply all the facts, so scientists must take the information and make educated guesses. Sometimes different scientists see the same material different ways, so there are many different theories.

Some paleontologists think the extinction was caused by a catastrophe, such as an asteroid's hitting the earth, or perhaps the erupting of a gigantic volcano. Others believe that a more gradual process was responsible. Some theories are that competition among dinosaurs and mammals was the cause; possibly, climate changes led to extinction.

Scientists also disagree about the amount of time it took for extinction to take place. Some think it happened in several days. Others say it took from hundreds of generations to more than half a million years. Any theory must account for the extinctions that occurred in the sea, including some types of clams and coiled mollusks.

The extinction event did not kill all animal and plant life. Many kinds of animals survived, including fishes, frogs, turtles, crocodilians, birds, and mammals. Scientists must take the fossil record and find reasons for all extinctions.

A duckbilled dinosaur watching an approaching comet, moments before the end of the Cretaceous Period.

The Cretaceous Extinction Event

What killed off the dinosaurs was a worldwide event. It affected many plant and animal groups, both on land and in water. Dinosaurs were only a small part—the disappearance of other living things was so widespread that scientists knew about the extinction 30 years before the first dinosaur was described.

Two duckbilled dinosaurs, following the comet's collision with Earth and the resulting fireball.

The victims of the Cretaceous extinction included dinosaurs, ammonites (mollusks related to the octopus and the chambered nautilus), pterosaurs (flying reptiles), and certain plant groups. But many other animal groups, even some reptile groups like champsosaurs, were not affected.

The image of the last majestic dinosaurs' passing away and leaving a world of shrewlike mammals and cold-blooded reptiles is false. Instead, many of the major modern land animals were already living in the Cretaceous. Dinosaurs shared their last million years with creatures we see today.

Fossil Record Information

The relationship between the plants and animals that became extinct at the end of the Cretaceous is important to the understanding of dinosaur extinction. The reasons for extinction in all these different groups are related. Understanding why one or another group became extinct may give us clues about dinosaur extinction.

Plants and some animals have a better fossil record than dinosaurs. They also give clues about climate that dinosaur fossils do not. Some small marine organisms called foraminifera have shells that tell the temperature of the water where the animal lived. By studying the chemical makeup of these tiny fossils, the temperatures of prehistoric seas can be found and changes in the ocean temperature can be shown. Ocean temperature reflects the climate and can be used to find out what the climate was like before and after dinosaurs became extinct and while extinction was occurring.

Edmontosaurus being pursued by *Tyrannosaurus*. Like all other dinosaurs living at the time, these two became extinct some 65 million years ago.

But unlike other scientists, paleontologists cannot go into a laboratory and repeat experiments. They can only untangle history by examining fossils. Unfortunately, information is often sketchy.

Scientists have studied rock formations containing the last dinosaur fossils and the boundary between the Cretaceous and the Tertiary, the period following the Mesozoic Era. In complete rock units, the best we can say is that dinosaur extinction took fewer than 100,000 years. In rock units where the sections are less complete, the time is longer—even as much as a million years.

In marine rocks—that is, rocks formed in seas or oceans—the record is often more nearly complete. From these rocks, the earliest and some of the strongest evidence has come for the extinction's having been caused by something extraterrestrial, or outside the earth's atmosphere. Because the record is more nearly complete in marine rocks, the time units studied are much smaller and more accurate. The time resolution (stratigraphic completeness) in rocks deposited in the San Juan Basin, a rock unit in western North America that includes the final phase of dinosaur history, is 100,000 years. In contrast, marine rocks deposited during this same time period in Spain allow an interval as short as 10,000 years to be studied. However, these rocks contain no dinosaur fossils.

A second problem with the fossil record is that sometimes fossils can be collected out of place. Recently, scientists reported that some dinosaurs may have survived the end of the Cretaceous and lived into the early Tertiary. Workers collected these specimens in rocks above the Cretaceous-Tertiary boundary, with specimens of early Tertiary mammals.

Corythosaurs against a sky suddenly afire in the aftermath of a meteor collision.

A group of *Triceratops* silhouetted against the evening sky. This large ceratopsian lived to the end of the Cretaceous.

Pachycephalosaurus, a dome-headed plant-eater from Late Cretaceous Montana.

Many paleontologists feel that these specimens were already fossils in the Early Tertiary. They believe that these fossils eroded out of the rocks, and nature reburied them along with the remains of Tertiary mammals. Scientists call this process "reworking," or secondary deposition. It only *looks* as if these animals lived at the same time: In fact, they lived millions of years apart. Paleontologists must study every specimen carefully.

Extinction Theories

There are two groups of extinction theories: catastrophic extinction and gradual extinction. Catastrophic extinction would have been caused by a sudden, external event, such as the collision of the earth with an asteroid or the eruption of a series of gigantic volcanoes. Gradual extinction would have been the result of changes in the earth's land mass and climate shifts. It could also have been because new and better-adapted animals won in the struggle for existence.

Until the recent theories about extraterrestrial collisions, some ideas about the disappearance of dinosaurs centered around mammals' beating them in the struggle to survive. One theory suggests that mammals killed dinosaurs by eating dinosaur eggs. Other scientists have suggested that dinosaurs caused their own extinction. According to this theory, too many meat-eating dinosaurs evolved, eating all the plant-eaters and therefore causing all dinosaurs to die. These ideas have the same pitfall described earlier. They explain dinosaur extinction, but ignore the extinction of other groups.

Tyrannosaurus rex.

183

Climate Change Theories

Leigh Van Valen and Robert Sloan have offered a more complex theory. During the Late Cretaceous, they say, the continents were moving and major new mountain chains began to rise. Many of the shallow Mesozoic seas dried up. The two scientists suggest that this caused the world's climate to change. Evidence from fossil plants at their Montana study site indicates that the average temperature dropped about ten degrees centigrade in the Late Cretaceous, a temperature decline that would have affected the earth. It would have become colder in the mountains, and new plants would have replaced existing warm-weather ones.

A *Maiasaura* herd smothering in an ash fall, following the Late Cretaceous eruption of several volcanoes nearby.

Van Valen and Sloan have argued that dinosaurs were at a disadvantage in the new forests of coniferous trees. So, they left the mountains and moved toward the tropics to a better climate. This theory suggests that the dinosaurs survived longer in the tropics than in the mountains. Because plants in the tropics survived, something else must have caused dinosaur extinction in these areas.

One suggestion is that placental mammals—mammals that give live birth and have a placenta—became abundant in the mountains because they no longer competed with dinosaurs. Later, the mammals escaped these regions and went into the tropics, where they drove dinosaurs to extinction. But there are problems with this theory. Some studies of foraminifera (the marine organisms talked about earlier) show that there was a short-term warming trend in the Late Cretaceous. But Van Valen and Sloan are right about the long-term cooling trend. There is no evidence

A *Triceratops* skeleton beside a stream. The dinosaur's Late Cretaceous environment looked strikingly contemporary.

that dinosaurs lasted longer in one area than another. Although competition between animals may explain the extinction of many living things at the end of the Cretaceous, other groups, especially the foraminifera, seem to have disappeared without competitors.

Another climate change theory comes from recent research. Some living reptiles—turtles and crocodiles—lay eggs. It was recently found that the sex of their offspring is decided by the temperature of the nest. So, if the same were true for dinosaurs, cooling temperatures in the Late Cretaceous may have caused all the young to be of the same sex, and the species would not have been able to reproduce and continue.

A *Dromaeosaurus* adult and juvenile eating a ceratopsid carcass. The predator was a relatively rare theropod—or, at least, rarely preserved.

Extraterrestrial Impact Theory

One exciting theory is that an extraterrestrial body, such as a comet or asteroid, hit the earth, causing the Late Cretaceous extinctions. Walter and Luis Alvarez and their coworkers found evidence for this during a study of some Cretaceous clay from northern Italy. To their surprise, they found that the clay was rich in the element iridium. Iridium is rare on earth and more common in extraterrestrial bodies, such as meteorites and comets. After further studies, the Alvarezes found the iridium only in a narrow layer. To their amazement, this iridium layer almost exactly matched the Cretaceous-Tertiary boundary. This led them to suggest that a large extraterrestrial body had hit the earth, causing the extinction of dinosaurs in the Cretaceous.

Two *Tyrannosaurus* witnessing a comet shower at the end of the Cretaceous.

Flying reptiles called pterosaurs silhouetted against the night sky following a meteor's collision with Earth. Pterosaurs, like dinosaurs, became extinct at the end of the Mesozoic Era.

Since this discovery, scientists have found the iridium layer, or "iridium datum plane," at the Cretaceous-Tertiary boundary at more than 50 sites worldwide. There is other evidence that a large body from outer space hit the earth. For instance, in an asteroid collision with Earth, intense heat and pressure develop. This causes changes in the rocks where the asteroid hit. One of these changes is shock-fractured quartz grains. Bits of quartz—a common mineral in the earth's crust—will break in an unusual way because of intense heat and pressure. The only other place shock-fractured quartz is found is at ground zero following explosion of an atomic bomb. Common elements also act differently when under intense pressure. For example, atmospheric nitrogen, a usually harmless gas, may have returned to earth as a deadly acid rain.

The impact of an asteroid would be a major event in the history of the earth. The iridium layer over the world shows that the extraterrestrial body that hit the earth must have been more than six miles wide. When it crashed, it would have been traveling at greater than 12 miles per second. The collision would have made a crater about 100 miles wide.

Because of its great speed, the asteroid would have ripped a giant hole in the earth's atmosphere, or the air covering the earth. Parts of the earth's crust would have been blown out into the upper atmosphere. Later, this would have rained back as tiny glass beads, ash, shock-fractured quartz, and parts of the asteroid.

A *Maiasaura* adult.

A large amount of dust would have covered the earth. The amount of dust caused by the 1815 explosion of Tambora, a volcano in Indonesia, caused climate changes worldwide for several years. The dust and debris that would have covered the earth following an asteroid hit of the size suggested by the Alvarezes would have been greater than any volcano.

The dust cloud would have taken from several weeks to many months to settle. First, the temperature on earth would have dropped below freezing because the dust clouds would have stopped the sun's rays from reaching the earth. This would have harmed the green plants and ocean plankton (plankton and green plants form the bottom of the world's food chain and also change carbon dioxide into breathable oxygen).

A herd of sauropods stampeding during a storm.

Late Cretaceous animals might have suffocated because of a lack of oxygen or have starved to death. This would have begun with the plant-eaters and carried through to the meat-eaters. After that, the dust cloud would have caused global warming, because the heat of the earth would have been trapped, unable to escape through the thick layer of dust in the upper atmosphere.

But soon after the Alvarezes announced the discovery of the iridium layer, other scientists began to find fault with it. Scientists have not determined where the asteroid landed. But if it were on the sea floor, it would be difficult to find. Scientists have proposed possible sites.

Tyrannosaurus bringing down the plant-eater *Edmontosaurus.* *Tyrannosaurus rex* was the largest known theropod on earth—and one of the last.

The dust cloud proposed by the Alvarezes would have killed all plant and animal life—not just dinosaurs. And paleontologists think the extinctions in the marine world lasted several thousand (perhaps up to 100,000) years. This may mean that the marine and land extinctions did not happen at the same time. This has led some scientists to suggest the possibility that many smaller meteors or comets hit the earth over a longer period of time.

Other Theories

There is a thin layer of carbon near the iridium layer. Some scientists think this may be because of global forest fires. These fires would have destroyed plants worldwide. They would have used oxygen and further heated the atmosphere.

Others scientists have suggested that gigantic volcanoes may have caused the iridium. A volcano this size has not erupted since the Late Cretaceous. There is more iridium inside the earth than in the earth's crust. So, large volcanoes that erupted from deep within the earth may have caused the iridium layer. They also may have caused the Cretaceous extinctions.

Finally, if we return just to the problem of dinosaur extinction, the proof of the sudden death of these animals is not great. In several dinosaur-bearing formations in the American West, workers have found the iridium layers, but have not found dinosaur fossils within ten feet of the iridium layer. If workers found no other fossils in these rocks, it would not be a problem. There could have been conditions that were not right for fossilization. However,

A drawing of a *Triceratops* head, showing its bisonlike "three-horned face" and curving neck frill. *Triceratops* was one of the last dinosaurs to walk the earth.

Three *Struthiomimus* escaping a forest fire caused by lava flow from a volcanic eruption. Iridium-layer studies suggest volcanoes, as well as asteroids and comets, could have been responsible for iridium deposits linked to an extinction event during the Late Cretaceous.

workers have collected fossils of mammals, crocodiles, fish, and plants in these layers. So, the extinction may really have taken several thousand years.

Some scientists think that comet showers causing extinction may be common. According to this theory, just as Halley's Comet passes the earth every 76 years, a swarm of comets passes the earth every 26 million years. These comets have their beginning in an area of protocomets called the Oort Cloud. Every 26 million years, the Death Star (a name given to a star that is, for now, only in scientists' theories) disturbs the comets, sending some toward Earth. These comets may hit the earth and begin a new chapter in the planet's history.

Conclusion

The Cretaceous extinction was not the first great extinction in the earth's history, nor would it be the last. The great Permian extinction happened millions of years before the first dinosaur lived.

A good explanation for dinosaur extinction does not just explain the information scientists have. It also makes predictions that can be tested and rejected if newer information favors another theory.

The causes for dinosaur extinction were complex. Even if an asteroid hit the earth and caused the extinction of dinosaurs, a changing climate or another factor also might have been involved. Even understanding the causes of recently extinct and endangered animals and plants is difficult. The problems with the fossil record, and in understanding conditions 65 million years ago, are huge. So, much of the dinosaur extinction debate is still theory.

Cretaceous uplands, with volcanoes. Some paleontologists suspect that volcanic eruptions during the latter part of the period may have led to the extinction of dinosaurs and other animal groups.

Tyrannosaurus rex, the "tyrant reptile king," at sunset.

INDEX

191